D1756530

WITHDRAWN

Vegetable Seed Production, 3rd Edition

Pershore College

VEGETABLE SEED PRODUCTION, 3RD EDITION

Raymond A.T. George

Former Senior Lecturer
University of Bath
UK

and

Consultant to the United Nations Food
and Agriculture Organization
Rome
Italy

www.cabi.org

CABI is a trading name of CAB International

CABI Head Office
Nosworthy Way
Wallingford
Oxfordshire OX10 8DE
UK

Tel: +44 (0)1491 832111
Fax: +44 (0)1491 833508
E-mail: cabi@cabi.org
Website: www.cabi.org

CABI North American Office
875 Massachusetts Avenue
7th Floor
Cambridge, MA 02139
USA

Tel: +1 617 395 4056
Fax: +1 617 354 6875
E-mail: cabi-nao@cabi.org

A catalogue record for this book is available from the British Library, London, UK.

First edition (published by Longman) 1985 ISBN: 978 0 58246 090 4
Second edition 1999 ISBN: 978 0 85199 336 2

Library of Congress Cataloging-in-Publication Data

George, Raymond A.T.
 Vegetable seed production / Raymond A.T. George. -- 3rd ed.
     p. cm.
 Includes bibliographical references and index.
 ISBN 978-1-84593-521-4 (alk. paper)
1. Vegetables--Seeds. 2. Food crop--Seeds. I. Title.

 SB324.75.G45 2010
 635'.0421--dc22

                              2009015985

ISBN: 978 1 84593 521 4

Typeset by SPi, Pondicherry, India.
Printed and bound in the UK by MPG Books Group, Bodmin.

The paper used for the text pages in this book is FSC certified. The FSC (Forest Stewardship Council) is an international network to promote responsible management of the world's forests.

Contents

Preface   vii

1 Organization   1

2 Principles of Seed Production   37

3 Agronomy   50

4 Harvesting and Processing   75

5 Storage   91

6 Seed Handling, Quality Control and Distribution   104

7 *Chenopodiaceae*   116

8 *Asteraceae* (formerly *Compositae*)   129

9 *Cruciferae*   140

10 *Cucurbitaceae*   162

11 *Leguminosae*   181

12 *Solanaceae*   202

13 *Apiaceae* (formerly *Umbelliferae*)   226

14 *Alliaceae*   251

15 *Gramineae*   264

16 *Amaranthaceae* and *Malvaceae*   270

Appendix 1   279

Appendix 2   281

References   283

General Index   299

Index of Species   315

Preface

The production of vegetables, their availability and supply continue to play a major role in the nutrition of the global population. Vegetables are essential for nutrition, and are vital dietary supplements to the basic staple food crops throughout the year. Therefore, the maximum potential for production of vegetables requires a stable and secure supply of high-quality vegetable seeds in order to maximize the production and quality of consumable produce. The timely availability of quality vegetable seed is imperative for all farmers, growers and gardeners to achieve satisfactory and successive yields wherever they cultivate any of this very important and diverse crop group.

This text is based on information and experiences acquired by the author while working in vegetable research and vegetable seed production projects in developing countries, and also experiences obtained from liaison with seed companies and seed-producing organizations in Africa, Asia, Europe, North America and the Pacific.

My observations and experiences gained from teaching and supervising undergraduates and postgraduates at university and college levels in the UK, in addition to organizing and teaching seed production training programmes for technical officers in developing countries, have been invaluable. I am very grateful to all the specialists, technicians, farmers and growers who have welcomed me to their enterprises and have also been so willing to share their experiences.

This third edition has included summaries at salient points within the first six chapters, dealing with the principles and practice of vegetable seed production. A suggested further reading list is provided at the conclusion of each of the 16 chapters. References to publications cited in the text follow Chapter 16. The general updating of this third edition has included more detail on 'organic' seed and its production, and also considerations on genetically modified organisms (GMOs). The increased links between vegetables produced from true botanical seed and those predominantly propagated vegetatively (i.e. the resulting planting material which is often referred to as 'seed') have been added to this edition in Appendix 1. The titles and numbers of UPOV's Test Guidelines that refer to vegetables are listed in Appendix 2 rather than including them in detail in the text.

My sincere thanks to A. Fenwickkelly and Peter R. Thoday for their support and useful discussions, especially during the proposal stage.

I thank Sarah Hulbert, Commissioning Editor at CAB International, for her patience and ever-ready advice while I was preparing the manuscript; and also Rachel Cutts, Associate Editor, and Kate Hill, Production Editor, for so kindly dealing with all the material which I generated for this third edition. The efficient and friendly attention to detail by Jaya Bharathi and colleagues at SPi, Pondicherry, India has been very much appreciated.

My thanks are also due to my family, Andrew, Christopher, Jane, Joseph and Patrick, who helped me out when major problems arose with the computer and who also improved my computer literacy from time to time.

Finally, my great appreciation to my wife, Audrey, who over the years has not only tolerated my absences on overseas missions, but also has been very patient while I have been at home preparing this third edition.

Raymond A.T. George

Bath
England
April 2009

1 Organization

The Role of Vegetables

Vegetable crop production is one of man's basic skills. Wherever humans have settled for long enough to produce a crop, they have cultivated vegetables for food and for animal fodder. The level of success and productivity originally depended on the local climate, seasons and the range of species cultivated. The cultivated species were developed by selection from local wild plants and subsequently supplemented by plant introductions from other areas, and later still from other continents. The story of cultivation has been described and discussed in interesting detail by Thoday (2007).

It is now being recognized that, in addition to their role as part of a balanced diet, vegetables can play a vital role in agricultural and social development processes. The Asian Vegetable Research and Development Center (AVRDC, 1998) has produced a plan that includes specific programmes such as the inclusion of vegetables in cereal-based systems and year-round vegetable production systems. The plan recognizes, and takes into account, the social and economic values of vegetables.

The pivotal role played by seeds in the development of civilization has been described by MacLeod (2007), who points out that seeds are a means of transferring genetic information from generation to generation, in addition to the essential nutrition and energy for germination, emergence and plant establishment. All of our vegetable species have developed as a result of generations of plantsmen selecting and growing the next generation from plants that have displayed the most desirable or useful characters or traits.

A successful vegetable production industry is very dependent upon a sustainable supply of satisfactory seeds. At the present time the seed industry plays an important role in both production and distribution of high-quality vegetable seeds. However, there are some communities where 'own seed' or 'on-farm' seed production is still the norm.

The production areas of vegetables range from large-scale farm enterprises and market gardens growing for profit, to private gardens, homesteads and subsistence farmers where vegetables are an essential element of the families' own efforts to supplement their diet or income. Figure 1.1 illustrates a demonstration vegetable garden at village level in the Shire Valley, Malawi; the demonstrations can provide, for example, information on nutrition, improved cultivars, techniques and indigenous landraces.

Vegetables are also cultivated in some communities for physical recreation or even for a pastime or hobby. There has been an increased interest in 'home-grown' and 'self-sufficiency' in some Western societies; this has arisen from the policy to increase physical activity as a way of improving health, partly following the global financial crises of 2008 and 2009, and in some cases including a desire to produce vegetables 'organically'. In some areas of less-developed countries, vegetables are not only grown for self-sufficiency but also for sale or exchange in village communities. The market growers in many areas have evolved from disposal of surplus crops to deliberate production for sale. With the further extension and development of urban communities and marketing systems, the commercial producer has continued to play an increasingly important role in meeting the vegetable requirements of the population.

Commercial production has extended considerably during recent decades in many parts of the world, as the field and protected cropping vegetable industries endeavour to provide continuity of supply for the fresh markets in urban areas (including supermarkets), the processing industry and export.

Fig. 1.1. A demonstration vegetable garden in the Shire Valley, Malawi.

Range of Vegetables

The range of morphological types of vegetable is very diverse and includes staple root crops such as potatoes and yams, leaf vegetables such as *Amaranthus* and lettuce, bulb crops such as onions and garlic, and edible fruit, which include tomatoes, peppers, squash and okra. Many of the legumes (i.e. members of the botanical family *Leguminosae*), e.g. *Phaseolus* spp. and *Cajanus cajan*, are not always regarded as vegetable crops, but when the immature seeds or pods are eaten (e.g. immature seeds of pea (*Pisum sativum*) and immature pods of green beans (*Phaseolus vulgaris*)), they are. When these are grown for large-scale production of the dried seed for processing or culinary purposes they are usually excluded from vegetable crop statistics and defined as 'grain legumes' or 'pulses'.

Classification of Vegetables

There are several systems of classifying the diverse vegetable crops grown throughout the world:

1. According to their culinary use or the part of the 'vegetable' plant that is consumed.
2. According to season or climatic area of production.
3. According to their botanical classification.

A classification according to their use is probably the best way of drawing attention to the diversity of genera and species globally cultivated as vegetables, but we do not gain any other useful information regarding their cultural or physiological requirements (see Table 1.1).

When vegetables are classified according to their season of production or climatic area, the diversity of species is apparent and there is some indication

Table 1.1. Examples of vegetables classified according to part of plant used as a vegetable.

Part consumed	Scientific name	Common name
Seedling	*Glycine max*	Soybean
Shoot	*Asparagus officinalis*	Asparagus
Leaf	*Amaranthus cruentus*	African spinach
Bud	*Brassica oleracea* var. gemmifera	Brussels sprout
Root	*Daucus carota* subsp. *sativus*	Carrot
Bulb	*Allium cepa*	Onion
Flower	*Cynara scolymus*	Globe artichoke
Fruit	*Lycopersicon lycopersicon*	Tomato
Seed	*Phaseolus vulgaris*	Haricot bean

Table 1.2. Examples of classification of vegetables according to season.

Scientific name	Common name
Warm season crops	
Cucurbita species	Melons, squash
Hibiscus esculentus	Okra
Zea mays	Sweetcorn
Cool season crops	
Apium graveolens var. *dulce*	Celery
Beta vulgaris L. subsp. *esculenta*	Beetroot, red beet

of each species' environmental requirements, although some ambiguities occur. This can also indicate the type of climate in which seed production may or may not be successful (see Table 1.2).

The most useful system for the classification of vegetables, when considering their seed production, is based on their taxonomy or botanical families. This draws attention to important botanical points such as the method of pollination and vernalization requirements, while at the same time agronomic aspects such as the need for crop rotation and specialized seed production techniques, e.g. harvesting method, seed extraction and seed processing, are still emphasized.

The botanical or taxonomic classification has therefore been adopted in the second part of this book for the main vegetable seed crops, with *Amarathaceae* and *Malvaceae* both in the final chapter.

The Importance of Vegetables

There are several reasons for growing vegetables, but the most important ones are that they are essential in the diet: they provide fibre, trace minerals, antioxidants, vitamins, folacin, carbohydrates and protein (Oomen and Grubben, 1978; Hollingsworth, 1981; Gormley, 1989). There is an increased emphasis in developing countries to supplement the staple foods (which include, for example, rice, maize or wheat) with locally produced green vegetables such as *Amaranthus* or Chinese cabbage. There is also a renewed emphasis on the identification of edible indigenous plant species, especially in developing countries where the local wild species have a high nutritious value but also fit into year-round cropping systems (Weinberger and Msuya, 2004). Figure 1.2 shows a trial and demonstration of indigenous vegetables in Arusha, Tanzania.

The medicinal and nutritional advantages of vegetables in the diet are receiving increased emphasis in the more developed areas of the world, for example Block (1985) has reviewed the chemistry of onions and garlic. National food advisors are encouraging the population of the UK to eat five different portions of vegetable and/or fruit a day (excluding potato) for a healthy diet.

Other reasons for an increasing interest in vegetable production include the use of leguminous crops in rotations to increase the soil's nitrogen status,

Fig. 1.2. Seed production specialists examining trials of indigenous vegetables in Arusha, Tanzania.

use of vegetables as break crops in arable farming systems, and the incentive to grow vegetables as high-value crops for the fresh market, for processing or export to earn foreign currency. Tan-Kim-Yong and Nikopornpun (1993) have described a scheme for the adoption of vegetable seed production as a substitute for opium poppy crops in northern Thailand; similar schemes are being explored in Afghanistan to find suitable high-value replacement horticultural crops for opium poppy.

An important reason why the demand for vegetables has increased in many areas of the world is the development and extension of a wider range of preservation techniques, such as canning, freezing and dehydration. This technology has led to a more diverse range of vegetable types and increased quantities being called for by the processors. Along with this there have been major improvements in the technology of large-scale vegetable production. The significant advances have included the adoption of herbicides suitable for individual crop species leading to crop establishment in a relatively weed-free environment, and the use of precision drilling and adoption of hybrid cultivars to improve crop uniformity and disease resistance. However, the adoption of some hybrid cultivars in which all the plant population matures at the same time is not necessarily an advantage for self-sufficient growers or subsistence farmers, unless the crop has been produced for processing or storage.

The increasing interest in vegetables, found at all levels or scales of production, has led to local, national and international activities to improve vegetable seed supply and quality. It can generally be stated that seed security leads to food security.

The Seed Industry and Seed Supply

There are two basic ways available for farmers to ensure a supply of seed for the production of the next generation of a vegetable crop. They are either saving seed from their own crop, which is usually referred to as 'on-farm seed production', or otherwise buying seed from elsewhere.

There are no figures available as to the percentage of vegetable crops in the world grown from farmers' own saved seed. In countries with well-developed agricultural and horticultural industries, such as the UK, France, Italy, the USA, Canada, Australia and New Zealand, the use of farm-saved vegetable seed is negligible, whereas in some developing countries as much as 90% of vegetable seed requirements are saved by farmers and growers from their own crops. It was reported at a workshop held to discuss the improvement of on-farm seed production for the Southern African Development Community (SADC) that farmers produce their own seeds of the supplementary staple crops, such as cowpea, millet and indigenous vegetables (Sikora, 1995). The proportion of 'on-farm seed' depends not only on the level of seed industry development in a country, but for some vegetable crops it also depends on how easily seed of suitable qualities can be produced in the local environment. The amount of extra work a grower is prepared to undertake to produce his own seed is also a factor. For example, it is relatively simple to save seed of the vegetable legumes such as peas and dwarf beans, but an extra season is normally needed to obtain seed from biennials such as onions. It can be argued that if vegetable growers produce their own seed, the plants for the seed crop have been subjected to selection pressures under the same environment as the market crop.

In addition to the management and economic factors associated with attempting to produce one's own vegetable seed, there are many difficulties resulting from the physiological requirements of some species in order to flower. For example, some species of the *Cruciferae* do not flower in the tropics where temperatures are never low enough for vernalization.

Seed Industry Development

Vegetable seed production within a country may be a public or private venture. In some countries, both public and private enterprises operate side by side. Joint ventures between a government and a private company are also possible; for example, the Government of Swaziland formed a joint venture with the US company, Pioneer Hi-Bred International, and the company is known as Swazi-American (PHI) Seeds Ltd (Wobil, 1994). The establishment of the East-West Seed Company resulting from a joint venture between Dutch capital and local expertise in The Philippines, Thailand and Indonesia has been described by Louwaars and van Marrewijk (1996).

In developing countries that do not have a vegetable seed industry, the feasibility of producing vegetable seed is most likely to be instigated by government initiative and, as the infrastructure develops, there is a transition of responsibility to the private sector. However, there is no hard and fast rule, and the organization of responsibility and overall development rate will depend largely on the priority given to vegetable crop improvement by individual national governments and the level of activity of the private sector. There is a general tendency for governments' interest in inputs aimed at vegetable crop improvement to follow in the wake of success with cereals or other nationally important staple crops.

There are several reasons why the initiative in vegetable seed production in a developing country is more likely to be taken by government than private enterprise. First, there may not be significant vegetable seed production expertise in the local private sector. Second, commercial companies within a country are unlikely to have the capital investment potential required for seed industry development, and even if they have, they may not be willing to invest in a commodity that requires the development of production units and also the necessary buildings and equipment for harvesting, processing, storage and distribution.

The need for seed industry growth in developing countries calls for a development rate far in excess of the relatively slow rate at which existing and well-established seed industries have evolved in countries such as the UK and the USA. An additional factor that sometimes calls for action by government rather than private enterprise in developing a seed industry is the country's socio-political beliefs and organization. A country that is isolated by its own or neighbouring countries' economic or social policies has a greater need to be self-sufficient.

Other factors that will have an effect on the rate of vegetable seed industry development include the emphasis put on vegetables and vegetable seed as a component of human nutrition, or their importance as a means of earning foreign exchange currency. These priorities are often determined by the relative parts of a government's policy for agricultural development; the country's national agricultural plan may even include a national seed policy (Gregg and Wannapee, 1985).

In many countries, there is already a well-developed and effective vegetable seed industry. Countries such as the UK and the USA have developed their seed industries over many years, while countries such as India, Kenya and Thailand have developed very successful seed industries in a relatively short time with considerable consolidation and development of infrastructure, production capabilities and marketing.

Seed Policies for Countries in Transmission

The phrase 'countries in transmission' refers to countries that are re-emerging as independent nations following political change. Examples include nations that were part of the former USSR where there is a transition from central control to a market economy. In the case of seed supply this usually entails a rebalancing of formal and informal seed sectors. Thus, more attention has been given in recent years to recovery programmes for seed industries in such countries. For example, this concept is being applied to states in Eastern Europe and elsewhere. (See Further Reading list at the end of Chapter 1.)

Assistance for Seed Industry Development

As the agriculture in a country develops, its government usually seeks to improve the vegetable seed situation. This often calls for financial and technical assistance.

There are many organizations that assist countries in the development of their seed industries. Some are capable of providing both financial and technical aid, e.g. the Food and Agriculture Organization of the United Nations (FAO), while others are in a position or willing to provide technical assistance for a seed project only, rather than finance its implementation. Conversely, there are organizations or donors with finance available for sound projects, but which do not have their own base for specialized technical assistance and advice. In the later case, the donor provides development assistance through an implementing or executing agency, such as FAO, a consultancy organization or consortium, which either has, or can recruit, the required expertise.

The Role of Seed in Agricultural and Horticultural Development

Most vegetable crops are grown from seed, although some major and minor crops are propagated vegetatively. A list of those crops that are entirely or mainly propagated vegetatively is given in Appendix 1.

It is important that growers, economists, planners, administrators and occasionally even politicians appreciate that seed is the starting point for the production of many vegetable crops. Other inputs, or investments, include items such as labour, fertilizers, irrigation, mechanization and crop protection materials (including fungicides, insecticides, acaricides, nematicides and herbicides). Many of these inputs depend directly or indirectly on fossil fuels, and it is increasingly vital that none of these important or valuable inputs is lost or wasted as a result of poor quality seed. In addition to the inputs listed above, there is competition for land for other purposes, including conservation, production of biofuels and demands from agricultural crops. It is therefore imperative that the seed used is of the best possible quality, regardless of the proportionate cost of seed to all other inputs.

Seed can, therefore, be identified as the primary and essential starting point of vegetable production programmes. This concept holds good for any scale of operation regardless of the size of the enterprise.

In many food programmes, the quantity and quality of available seed will have a direct effect on the amounts of fresh vegetables available. The range of types and available quantities of each of them over as long a season as possible will be affected by availability of seed of suitable cultivars for succession.

New cultivars that are either developed within a country or introduced from elsewhere have to be distributed through the mechanism for seed supply. Many modern cultivars are relatively high yielding compared with their predecessors. There are cultivars that have been bred for resistance to specific pests or pathogens, for example tomato cultivars with resistance to specific nematodes and also *Fusarium* and *Verticillium* wilts. Some have been developed by plant breeders for specific seasons or purposes, for example lettuce that is suitable for greenhouse production during winter, or 'stringless' dwarf beans for processing, and self-blanching celery for winter field production around the Mediterranean to be marketed in northern Europe. In other crops, such as cabbage, F1 cultivars with a high degree of uniformity have been developed for increased yield and heat tolerance. The AVRDC has developed breeding material of soybean, which has been further developed in Thailand and released as a vegetable cultivar (Chotiyarnwong *et al.*, 1996).

New vegetable crops to widen the range of available produce are introduced into areas through the seed trade's marketing or distribution network within a country or state. In this way new crops (e.g. Chinese cabbage introduced into the protected cropping system of northern Europe) or new methods of preservation and presentation with existing crops (such as processed peas, frozen peas, canned tomatoes, tomato paste, onions for dehydration and cabbage for coleslaw) have been introduced into the horticultural industry.

Some of the technological developments made by agricultural and horticultural research institutes have been adopted by seed companies. Examples include seed treatments for the control of specific diseases, such as the technique for the application of thiram as a seed soak (Maude *et al.*, 1969). Similarly, grading and pelleting of seed lots to facilitate their use by growers in conjunction with precision drills or machines for sowing in soil blocks to raise transplants are done by the seed trade or its related agencies.

The possibilities of including crop protection chemicals, growth regulators, materials to accelerate germination under dry conditions and nutrients in materials used in seed pelleting are constantly under review. As a result of the increased interest in environmental issues, there is an increased incentive to find techniques that use less pesticide per unit area of field or plot. Smaller amounts of crop protection chemicals applied to seed before sowing are considered by some agronomists to be more 'environmentally friendly' than postemergence treatments, which generally apply more chemical per unit area, and can also be considered safer for field workers and operators.

Positive steps, as research progress is made, are taken up by seed companies for the benefit of farmers, growers and gardeners; many of the larger or specialist seed companies have their own research departments working on seed-related issues or have business links to pesticide or seed treatment companies.

In addition to the role of vegetables in improving the standard of living by way of diet and health, vegetable crops are a part of agricultural systems. Any improvement in agricultural income and contribution to profitability from high-value horticultural crops such as vegetables and fruit crops will play an important role in improving a country's economy.

There are, therefore, many indirect and direct ways in which a dynamic vegetable seed industry can assist a country, regardless of the development stage that the country has reached.

Summary of the Role of Vegetables and Satisfactory Seed Supply

- Vegetables provide essential components in the diet, including dietary fibre, trace minerals, vitamin C and folacin.
- Reasons for production include substance farming, local trading, commercial growing for fresh market and processing, healthy lifestyle (e.g. nutrition and physical well-being), competition or interest in organic lifestyle. Classification of vegetables is carried out according to culinary use or morphological part consumed, by season or climatic area for production, and according to botanical classification.
- Seeds are obtained from saving one's own ('on-farm'), national government seed programme, commercial suppliers or aid programmes; also combinations of any of these sources.

Assistance for Seed Industry Development

There are several sources of assistance for seed programmes in developing countries. Some of the assistance may be financial; the conditions imposed by the donor or lending organization may depend on the level of agricultural development, or economic or political stability in the recipient country. The following types of organizations may be involved in financial assistance for the seed industry development.

The United Nations agencies

The United Nations (UN) generally finances seed programmes via the United Nations Development Programme (UNDP). Occasionally special funding may be available from the UN Food and Agriculture Organization (FAO). A major role within the UN system has been played by the FAO Seed Improvement and Development Programme (SIDP), described by Feistritzer (1981).

Development banks

These include the World Bank, the Asian Development Bank and the Bank for Economic Arab Development in Africa. These organizations usually have their own expert staff or consultants who assess requirements; traditionally the banks have generally made loans at fixed rates of interest with repayment over a relatively long period, although with major changes going on with financial policies the loan conditions may vary.

Bloc aid

This type of funding can be within political groups of countries or from a group of countries formed for trading and other purposes. Examples of such groups who have provided aid for seed-related projects include the European Economic

Community (EEC), the Near East Governments' Co-operative Programme and the Arab Organization for Agricultural Development (AOAD).

Bilateral aid

This is the direct financial and/or technical assistance from one country to another. It is not necessarily between adjacent countries, for example a European country may well provide bilateral aid to an Asian country, for example UK to India, Germany to Tanzania or Switzerland to Nepal.

Bilateral aid is often administered through an official organization of the donor country. Examples of these include the Swedish International Development Agency (SIDA), the Danish International Development Agency (DANIDA) and the German Deutsche Gesellschaft für Technische Zusammenarbeit (GTZ). Other forms of financial assistance or aid may be given by organizations such as the United States Agency for International Development (USAID) and the Ford and Rockefeller Foundations.

Non-government organizations

The non-government organizations (NGOs) are very active in developing countries. They usually operate on a charitable basis for relief and development work. Although collectively they cover many areas of the world, very few have an obligation solely to seed production or seed industry development. Those which include agronomy in their remit are more usually operating in projects confined to relatively small geographical areas, even if the NGO operates regionally or globally. Typical involvements of NGOs in seed projects have been described by Muscapole (1995) and Wilson (1995).

Role of the recipient country

The extent to which a country is prepared to provide financial assistance for its own seed industry development is often seen by donor or loan agencies as a measure of the country's intent. Whereas financial inputs from outside the country are sometimes the only way of getting a proposed seed scheme started, it is the host country that must maintain and further consolidate the infrastructure to the extent that it is sustainable, once the initial project has come to an end.

Sources of technical assistance for seed industry development

The initial teaching, training and technical advice relating to the development of a seed industry is usually provided by specialists from countries that have already achieved a satisfactory seed industry status. The specialists are either in full-time employment with the donor organizations or are released from their normal duties on short-, medium- or long-term loan. Thus, specialists can be made available from national government departments such as ministries of agriculture or educational establishments such as universities or specialist extension services. It is a usual practice for the specialist to work with counterpart personnel in the recipient country. The availability of local counterparts is vital because the general concept is that when the development project has been completed, for example after a period of 3 years, the local counterparts will continue more or less unaided, having in the meantime become more experienced and conversant in the required technology. However, the longer-term success will depend on sustained national inputs and support. The transfer of information and development of skills at all levels is frequently referred to as 'transfer of technology'. In addition to the different national and international agencies mentioned earlier, there are national and international consultant organizations, which employ their own specialist teams and are engaged by a government or other organization on behalf of a developing country to organize and execute a specific development project. These consultant organizations frequently compete with each other to secure a specific contract by submitting tenders with outline programmes.

Summary of Assistance for Seed Industry Development

- Existence of a government's national seed policy is usually regarded as an essential starting point for further development.
- Assistance should be considered in the light of nutrition, and agricultural and social development.
- There is a need for national government to meet potential donors part way (this does not refer to seed relief following disasters).
- There is a need for acceptable programme proposal for recipient government (national, federal or regional) and the sponsor (donor).
- There is a need for implementing organization to have suitable technologists available and recipient to have its agreed inputs, including suitable counterparts available for training.
- There should be a provision for different sources and types of assistance for training, examples include in- or out-of-country training, regional training courses and one-to-one with counterparts.

Assessment of Vegetable Seed Programme Requirements

An assessment of the vegetable seed situation in a specific country is essential before recommendations for seed industry development can be made. It is, therefore, very important to obtain information, data and records of observations from a wide area and to consult as many sources as possible within the country in the search for information and evidence of needs.

The assessment should first study the general vegetable crop situation to determine which constraints result directly from insufficient seed of satisfactory quality. It should include observations on agronomic practices such as sowing rates, cultivar suitability and presence of seed-borne pests and pathogens in the growing crops.

Having obtained an overview of the current vegetable production, the different facets of the prevailing seed industry and seed supply should be evaluated. All aspects of the development of seed programmes have been dealt with very comprehensibly by Douglas (1980).

The points that should be taken into account regarding seed sources include material from the public and private sectors, and grower's own saved seed. The particular points relating to the problems of vegetable seed production have been discussed by George (1978). A systematic procedure will ensure that sufficient background information is obtained so that recommendations for comprehensive vegetable seed projects or for the improvement of any part of vegetable seed production, supply or quality control can be based on the findings of the relevant parts of the framework.

The following points should be taken into account before attempting to improve the vegetable seed situation:

1. Background: the relationship of agricultural and vegetable crop production to the overall national agricultural policy for development. The relationship between the vegetable seed programme to the overall national seed policy and seed programme.

2. Vegetable crops produced: their location, relative importance and market trends; possible changes in emphasis or importance of specific crops; interaction of vegetable crops with other agricultural crops or animal husbandry groups; effectiveness of research or extension services on improved technology; sowing rates and estimated annual seed requirements of individual crops, including crops for specific seasons or market outlets.

3. Cultivars: the range of cultivars available, their suitability for local environmental conditions, production systems, market requirements, market outlets and consumer preference; origins of cultivars (i.e. indigenous, local or imported), rate of replacement; existence, scope and efficiency of cultivar trials.

4. Seed production: the seed industry status and the vegetable seed production relative position, status and development. Quality and development stage of any existing vegetable seed production. Sources of vegetable seed used by growers. Identification of existing or potential, vegetable seed production areas. Level of husbandry and technology used in current vegetable seed production; current seed yields of each species; existence of distinct seed classes.

5. Seed processing: level of development of seed processing and degree of mechanization for harvesting, threshing and cleaning.

6. Seed storage: adequate storage technology in use; effects of local conditions on vegetable seed quality and quantity. Adequate reserves for contingencies. Effectiveness of stock control, records of each seed lot's history; an efficient labelling system.

7. Seed marketing: the extent to which the current vegetable seed supply is meeting demands in all areas and types of production. Adequate promotion of vegetable seed and the range available down to village level, role of extension services and demonstration of improved vegetable seed.

8. Legislation: existing legislation for other crop groups; inclusion of vegetable seed in existing schemes for quality control; germination or other such testing services or quality assurance, including certification schemes.

9. Policy: existing national and/or regional government policy affecting vegetable seed requirements; adequate finance and manpower available for establishment or further development of requirements. Possible effects of changes in government policy regarding role of the private sector. Existence of training schemes and facilities for all levels of personnel of both genders, including technicians, seed producers, vegetable growers and farmers.

Summary of Points to Observe When Assessing Seed Programme Requirements

- Background of vegetable crop production.
- Vegetable crops produced and their relative importance.
- Cultivars, local varieties and landraces of importance.
- Seed production, level of development.
- Seed processing, level of development.
- Seed storage, including provision for contingencies.
- Seed marketing.
- Legislation and its effectiveness.
- National or regional seed policies.
- Different sources and types of assistance for training, examples include in- or out-of-country training, regional training courses and one to one with counterparts.

The roles of government and private enterprise in the supply and distribution of seed have been described and discussed by Kelly (1994).

Seed as an International Commodity

Since the development of specialist seed production areas in different parts of the world, there has been a tendency for companies or organizations to specialize in the multiplication and distribution of seed. Sgaravatti and Beaney (1996) have produced a world list of seed sources, which includes private and public sector institutions producing seed (including vegetable seed) in over 150 countries.

The actual quantity of seed produced by an individual organization will depend on several factors including the crop species dealt with, cultivars' multiplication rates (largely genetically controlled), seasonal conditions, husbandry and skills of managers, growers and technicians. Many of these companies are based in countries such as the USA, the UK, the Netherlands, France, Australia, New Zealand and South Africa. A large proportion of the seed produced by such companies is produced abroad, often in a different region of the world. A seed company based in the UK, for example, is likely to be producing the same crop species, even the same cultivar in more than one area of the world simultaneously. Pea seed, for example, may be produced in Hungary and New Zealand by the same company at different times of the year and using different contractors. A company will usually supply stock or basic seed to an overseas farmer (or contractor) on the basis of buying back from him all that is produced, but paying him according to the quality and purity of the seed lot produced. A detailed account of the different schemes used for contract seed production has been described by Chopra and Chopra (1998). Hossain (1996) has described vegetable seed production by contract growers.

Seed companies continue to look for new production areas and new markets where not only climatic conditions but also the economic situation, including transfer of capital investment and local labour costs, are favourable. The cost of labour is an important consideration in the production of hybrid seed, especially when hand emasculation and pollination are necessary. For these reasons there has been increased activity by seed companies to establish production contracts in countries such as India, Indonesia, Pakistan and Taiwan.

At one time, seed of some crops was produced in commercial quantities on the off-chance that it would be purchased by an entrepreneur and subsequently resold at a profit to other organizations in the distribution chain. There is considerably less of this type of activity now, especially with vegetable seed, and most of the production is planned according to short- or long-term stock requirements, based on specialist knowledge of the trade. Despite this stability, seed companies continue to look for new markets where sales potential is likely to increase (e.g. the development of large-scale vegetable production in a developing country).

This stabilization and organization of the seed industry have probably been major contributors in the advancement and introduction of legislation leading to quality control.

All the international activity and trade means that seed can be subject to price fluctuations or inflation resulting from monetary changes, for example international exchange rates and any premiums on foreign exchange. The existence of tariffs or similar trade barriers will affect price and possibly the choice of location for production. Other economic hazards or factors that can have an effect include trade embargoes or sanctions.

Examples of Organizations Operating Internationally to Assist Seed Trading and Policies

The International Seed Federation

The International Seed Federation (Fédération Internationale du Commerce des Semences (ISF)) was formed in 1924 to discuss common problems and interests. At that stage of world seed industry development there was relatively little coordinated international activity. A history of the Federation was published to mark its 50th anniversary (ISF, 1974). By 2008, there were 70 developed and developing member countries covering all continents.

In addition to dealing with trading problems with which the seed industry is faced from time to time, ISF advises on the settlement of international disputes according to its own arbitration procedures agreed upon, and accepted by, members (ISF, 1996a). ISF has sets of rules for each of the main crop groups, including vegetables (ISF, 1994). These rules deal with such items or aspects of the international trade as contracts relating to offers and sales, import and export licences and definitions of trading terms. Seed quality is also defined in relation to trade, and the information required to be given with a seed lot is listed. Other aspects included are business requirements such as shipping, insurance and packing agreements, and there is a formula for calculating compensations relating to disputes. The organization sees its own rules as being more realistic and appropriate to its arbitration procedures than international law, which can be time consuming and detrimental to satisfactory settlement within the life of a given seed stock.

The ISF cooperates with other international organizations involved with seed, including the International Seed Testing Association (ISTA) and the Organisation for Economic Co-operation and Development (OECD). Le Buanec (1998) has described the objectives, background, legal framework and achievements of ISF in detail. The Federation's current strategic plan is described in a ISF document (ISF, 1996b).

Organisation for Economic Co-operation and Development

The OECD is an intergovernmental organization involved with economic development, employment and expansion of world trade. It has a Directorate for Food, Agriculture and Fisheries, which administers the OECD Seed Schemes; these schemes include rules and directions for seed moving internationally and also the extension of seed certification. The OECD Scheme for Certification of Vegetable Seed was first introduced in 1971 and is frequently revised and updated (OECD, 1996). OECD regularly publishes a list of cultivars eligible for certification (OECD, 1995a). National seed certification schemes usually follow the OECD scheme exactly or otherwise are very frequently based on it. Bowring (1998) has described the objectives, background, legal framework, achievements and future developments of OECD Seed Schemes. The concept and components of seed certification are discussed later in this chapter.

Asia and Pacific Seed Association

The Asia and Pacific Seed Association (APSA) was founded in 1994. The Association is composed of a wide range of seed enterprises and seed organizations in the region. It has as its main aim the improved production and trade of agricultural and horticultural seed and planting material in the Asia and Pacific region. APSA is an international, non-profit and non-governmental association with a regional forum for the encouragement of collaboration among seed enterprises. The Association's activities include the representation of members' interests to governments; also compiling and disseminating information on technical, regulatory and marketing issues. In addition, it assists in the organization of training and cultivar testing programmes. The Association produces a bimonthly publication, *Asian Seed*. The aims of APSA have been described by Lemonious (1998).

Summary of Assistance to Seed Trade and Legislation

- Seed can be recognized as an international commodity.
- Several organizations assist with the commercial, technical and forms of protection for producer, supplier and consumer.
- Examples of these include ISF, OECD, UPOV and regional organizations (e.g. APSA and SADC).

The Role of Plant Breeders in Seed Industry Development

The science of genetics and plant breeding has been applied to the maintenance and production of vegetable cultivars only since the beginning of the 20th century. Even at the present time, especially in developing countries, there are many cultivars maintained by growers who simply save seed from selected or unselected plants.

A good example of early cultivar development by people with an 'eye for a good plant' rather than scientific breeding is the diversity of cultivars of the *Brassica oleracea* subspecies, which includes Brussels sprouts, cabbages and cauliflowers. These were selected and maintained by individuals working in isolation and often with no biological training. These selections by gardeners and growers have resulted in the wide range of types within each subspecies. Examples include cabbages for successive seasons with relatively different morphology, cauliflowers suitable for specific production areas and the different types of sprouting broccoli. The history and development of these cole crops was discussed by Nieuwhof (1969), who reviewed the important gardeners' and growers' skills that produced the present wealth of germ plasm within the species. The application of inbreeding and production of commercially available F1 hybrids have been applied to cole crops such as Brussels sprouts and cabbages only since the late 1950s or so.

The diversity within other vegetable crop species can also be attributed to the selection process of successive generations of gardeners and commercial vegetable growers. The wide range of melons and other species in *Cucurbitaceae* are examples of this form of selection in arid and tropical areas of the world.

In more recent times, modern plant breeding methods have been applied to the development of vegetable cultivars. These have resulted in attributes such as increased yield, improved quality, extension of season and resistance to

specific pests and pathogens. Innes (1983) reviewed the breeding of field vegetables. The progress that has been achieved in the breeding of F1 hybrid
vegetable cultivars has been reviewed by Riggs (1987).

Plant breeding programmes can be initiated in the public and private sectors, and in many countries they exist simultaneously in both. In some cases
there is close liaison between national institutes and private companies. Public
sector stations may work on plant breeding and genetic problems, which would
require too high an investment of time and money to be undertaken by private
or commercial workers. In addition to the development of breeding techniques,
public sector institutes may demonstrate the application of a particular technique or breeding method by producing a new cultivar as the end product.
There has been a call for more direct liaison between plant breeders and farmers in developing countries in order to ensure that the end product has attributes
acceptable to the consumer (Sperling *et al.*, 1993).

The leading international institute for the development of vegetable breeding material, especially for the tropics and subtropics, is the AVRDC, which has
its main station in Taiwan. It has an Asian Regional Center in Thailand, an
African Regional Programme operating from Arusha, Tanzania, and project
offices in Bangladesh and Costa Rica. The crop improvement programme of
AVRDC for the major vegetable crop species grown in the hot humid tropics
has been discussed by Opeña *et al.* (1987).

AVRDC has guiding principles on plant genetic resources and developed
plant material, these are in line with the articles adopted by the UN Conference
on the Environment and Development (UNECD). The AVRDC makes Plant
Material Transfer Agreements specific for the release of 'plant materials' to the
public sector and 'improved materials' to the public and private sectors.

The Role of the International Plant Genetic Resources Institute (IPGRI)

The national and international interest in cultivar development has promoted
attention to the need to collect, evaluate and preserve germplasm. The International Plant Genetic Resources Institute (IPGRI) is active in identifying,
describing and promoting national and international germplasm collections of
a range of species including the numerous and diverse vegetable species in the
world. This international institute has described details of its strategy, plans and
their implementation in a policy document (IPGRI, 1993).

The Institute has five regional groups, with collective global coverage.
IPGRI's main focus is on the developing countries, but the regional groups also
establish contacts with institutions in the more developed countries.

Plant Breeders' Rights

Plant breeders, or their employers, rely very heavily on income from the sale of
seed (or planting material) of the cultivars that they have developed. This concept is especially important for the private sector; however, the public sector is

often very dependent on income from new material so as to recoup development and production costs and to receive income for further research and development. Thus, the granting of plant breeders' rights to the breeder of a new cultivar provides the breeder, or the breeder's institute, protection from the multiplication or reproduction of the cultivar by other unauthorized persons. The registration of a new cultivar and the description of its distinct characters with a registration authority, which is recognized by law, protects the breeder from unlawful reproduction of the material, or cultivar, by persons in countries that have adopted the system of plant breeders' rights.

The relevant aspects of measures and associated legislation for the protection of the plant breeder have been discussed by Kelly (1994). The Convention of the International Union for the Protection of New Varieties of Plants (UPOV) has been set up to ensure that plant breeders can have exclusive property rights for their new cultivars. A comprehensive account of UPOV and its functions are given by Greengrass (1998). It should be noted that UPOV uses 'variety' rather than 'cultivar' in its documents. The Organization's main publication for information on the development of plant cultivar protection legislation in the world is *Plant Variety Protection* (the UPOV Gazette and Newsletter), which is published several times each year. The main criteria for a cultivar to be included for protection are that it is distinct from existing, commonly known cultivars, sufficiently homogeneous, stable and new (i.e. it has not been commercialized before certain dates established by reference to the date of application for protection). UPOV Publications of test guidelines relating to vegetable cultivar descriptions are listed in Appendix 2.

Plant Nomenclature

The term 'variety' has traditionally been given to an assemblage of plants of the same species, which have the same characters distinguishing them from other assemblages of the species. The rules for plant nomenclature were studied in the 1950s by a working group, and their initial considerations and code were later reviewed and revised. The code has been accepted by different countries as a basis for legislation relating to plant nomenclature. The terms 'cultivar' and 'variety' are taken as exact equivalents, although many national and international acts and regulations retain the word 'variety' rather than 'cultivar'. According to Article 10 of the International Code of Cultivated Plants (Anonymous, 1980a) the international term for 'cultivar' denotes an assemblage of plants that is clearly distinguished by any characters (morphological, physiological, cytological, chemical or others) and which, when reproduced (sexually or asexually), retains its distinguishing characters. The cultivar is the lowest category under which names are recognized in this code. The term 'cultivar' is derived from cultivated variety, or their etymological equivalents in other languages. Article 10 notes that the concept of cultivar is essentially different from the concept of botanical variety, varietas, which is a category below that of species.

All vegetables in cultivation are clearly named according to these principles. Each has a generic name, a specific name and a cultivar name. For example, if we are referring to a certain cultivated variety (i.e. cultivar) of onion known to seedsmen and growers as 'Texas Early Grano', it is identified as *Allium cepa* cv. 'Texas Early Grano'. In this example, *Allium* is the generic name, *cepa* is the specific name (sometimes referred to by taxonomists as the specific epithet) and 'Texas Early Grano' is the cultivar name. Quite frequently, in order to ensure the exact species that is being referred to, we give an 'authority' following the specific or species name. In the case of *Allium cepa*, the authority (i.e. the botanist who first described it and named the species) was Linnaeus. Thus, an abbreviation of Linnaeus is added after the species name, and it would appear in print as 'L.'. Different types of the same species, as occur for example in *Beta vulgaris*, are referred to as subspecies, abbreviated to subsp., for example the scientific name for beetroot is *Beta vulgaris* L. subsp. *esculenta*. Where the taxon is smaller than a subspecies, but includes several groups, each may be referred to as a convarietas, which is abbreviated to convar.; as an example, some workers refer to the different types of *Brassica oleracea* as convars., e.g. cabbage as *Brassica oleracea* convar. *Capitata*. For other examples of species in which convar. is used and a glossary of common names see ISTA (2007).

There are some variations in nomenclature, for example within the EEC separate species names are still used for field pea (*Pisum arvense* L.) and garden pea (*Pisum sativum* L.) with retention of the original specific epithets given by Linnaeus. However, the various types derived from the two species are now considered to be too closely related and all of the cultivated types are referred to as *Pisum sativum* L. *sensu lato* (i.e. 'in the broadest sense').

The ISTA has produced a List of Stabilized Plant Names (ISTA, 1988) and, as far as possible, these names are used in this book. From Chapter 7 onwards the authorities will only be given when the scientific name of a plant occurs for the first time.

Cultivar names

The International Code of Botanical Nomenclature is not enforced or upheld internationally by legislation but many governments voluntarily subscribe to its success by adopting its recommendations, which are generally accepted by plant breeders. A plant breeder, or national breeding institute, will not normally give a cultivar name for a new line until the material has been seen to perform well in national cultivar trials and is likely to be acceptable on a national list. Until this is clear, the material is entered into trials under a code number or name. This avoids subsequent problems arising if the material is either not acceptable as distinct from, or has no clear advantages over, existing cultivars. Thus, the proposed name, which may have a seed house's or breeding institute's prefix, is not associated with inadequate material. Similarly, if a breeder has a special name in mind, it is not wasted on what may be considered subsequently as deleterious material.

The International Code of Nomenclature lists the aspects of cultivar names that should be avoided; the main points relating to vegetables and that should be excluded are:

1. Arbitrary succession of letters, abbreviations or numbers.
2. An initial article, unless it is the linguistic custom.
3. Names commencing with an abbreviation.
4. Names containing a form of address.
5. Names containing excessively long words or phrases.
6. Exaggeration of the qualities of the material.
7. Names that are attributed to other cultivars.
8. Names that may be confused with existing cultivar names.
9. Inclusion of words such as 'cross' or 'hybrid'.
10. Names exceeding three words (where an arbitrary sequence of letters, number or abbreviation is counted as one word).
11. Latinized names.

Cultivar Release

Traditionally, growers, farmers and gardeners have become aware of new cultivars via the range of lists, catalogues and other media of the commercial seed companies. In many countries, especially developing countries, the extension services have taken a lead in making farmers and growers aware of improved cultivars and although this practice still prevails, legislation in some countries requires that cultivars of some crop species are evaluated at national level to ensure that new cultivars are released into commerce, only if they are clearly distinct from those already on the market. This information is also used before cultivars are selected for inclusion in national lists of recommended cultivars.

There are several advantages in this system including avoiding synonyms, providing information necessary for the effective enforcement of plant breeders' rights, providing the consumer with information on new cultivars with their cropping performance and morphological characters.

There has been, especially during the 1950s, a concerted effort by interested parties, including the seed traders', breeders' and growers' organizations, to reduce the number of synonyms of vegetable crops. For example, the existence of synonyms for lettuce types was demonstrated by Watts (1955), for spring cabbage by Johnson (1956), for red beet by Holland (1957) and for forcing radish by Watts and George (1958). This type of study was usually made as a prelude to a proposed breeding programme in order to assess the existing material available to growers.

There were two main reasons why synonyms such as those cited by these authors had come into being. First, many synonyms had become popular because different lots of seed material from the same original stock were taken and multiplied by different commercial seed organizations over successive plant generations. These separate lines remained sufficiently similar to be regarded as synonyms. Second, many seed companies preferred their own cultivar name for a particular genotype and were offering the same seed stock under different cultivar names even though they had each obtained it from the same primary source. This practice was not illegal at that time and the introduction of legislation to avoid this situation occurring is relatively recent in many countries. Many customers of seed companies had preference for a seed stock with a cultivar name associated with a particular company; this provides a very good and early example of customer loyalty to their favoured seed supplier.

With the introduction of plant breeders' rights and the formation of national lists it has become necessary to ensure that any new material released on the market is in fact clearly different from existing material. The distinct characters of such new material are used to distinguish the cultivar for registration under plant breeders' rights. This distinctness, plus proven uniformity of the cultivar and its stability for these characters when multiplied over several generations, is now examined and tested in official trials necessary for the 'lists of recommended cultivars' compiled by official organizations within an individual country, or occasionally groups of countries. Where several countries in close geographical proximity have trading or other agreements, there is now a tendency for them to compile group lists of cultivars for their community. The lists issued by the EEC are an example of this trend.

Distinctness, uniformity and stability tests

The evaluation of cultivars for distinctness, uniformity and stability (DUS) enables controlling authorities to regulate the release of cultivars via national lists and control plant breeders' rights.

It is recognized that while a particular cultivar is a relatively unique genotype, there is first the possibility that the material may not be completely homozygous. For example, whereas an F1 hybrid would be expected to be composed of a uniform population, an open-pollinated cultivar of a species that is largely cross-pollinated will display a degree of difference between individuals in the population. In addition, successive multiplications will put different selection pressures on the cultivar. These selection pressures will depend on a range of factors including the pollination system of the species (i.e. whether predominantly self- or cross-pollinated), the climate and environment where the multiplication takes place, and the criteria used by the person selecting plants to produce the next generation. The criteria used for DUS testing depend on the species but several characters such as morphology, time of flowering and resistance to specific pathogens are useful where they are less likely to be affected by environment.

Cultivar trials

There are several reasons for conducting field trials for the evaluation of cultivars belonging to the same species that are in addition to those organized specifically for cultivar registration or DUS testing. Plant breeders may sometimes conduct field trials to examine all the available material as a prelude to a specific breeding programme. The results of this type of plant breeders' screening trials are not always published, but a typical example was described by Fennel and Dowker (1979), who screened onion material for characteristics required for a breeding programme, which hoped to improve onions suitable for autumn sowing in the UK.

The type of cultivar trial discussed here is that which in broad terms assists farmers and growers to select from a list those cultivars that are suitable for specific seasons, production systems, maturity periods and market outlets. This range of criteria applies equally well to the decision making of large-scale commercial producers and subsistence farmers. This type of trial is usually referred to as a 'variety trial', although the term 'cultivar trial' is also referred to. An early example of a variety trial to evaluate celery cultivars was done by the Royal Horticultural Society in its gardens at Chiswick in 1874, when it trialled and tested celery to select varieties suitable for production in trenches. Its assessments included the more desirable vertical growth habit in addition to leaf stem colour (Shaw, 1889). A well-organized lettuce cultivar trial in the Netherlands is illustrated in Fig. 1.3.

It is important that the objectives of a cultivar trial are clearly defined at the outset. All too often, especially in developing countries, the so-called variety trials are simply a cursory look at the plants grown from free seed samples, randomly and indiscriminately obtained from the seed trade, or agencies willing to provide samples for trial purposes. This type of 'screening' of material can be sometimes useful as a 'growing-out test' in a single season in order to obtain a general impression of the individual samples before deciding whether or not to include them in larger-scale trials. It is normal practice for the suppliers of samples to request some feedback or information about the performance of each

Fig. 1.3. Trial of outdoor lettuce cultivars in the Netherlands.

sample, unfortunately they do not always receive constructive comments from the recipients.

The results and information obtained from a well-organized cultivar trial should provide information for farmers and growers on the suitability of cultivars for specific purposes. If the trials indicate a wide range of deficiencies in the available material, it may be necessary to search further afield for suitable cultivars or alternatively to undertake subsequent breeding programmes.

Cultivar trials can provide information on where priorities should be placed in relation to national seed production or importation, and are essential in developing countries when deciding on a seed programme. Lists can be compiled from results of trials to help farmers, extension workers or advisers, and other interested parties. Kelly (in Feistritzer and Kelly, 1978) defined three types of lists that can result from cultivar trials, and each contains 'preferred' cultivars (i.e. cultivars that are better than those not listed). The lists are of three kinds, i.e. descriptive, recommended or restrictive.

Descriptive cultivar lists

This type of list aims to assist farmers and growers to choose from the available cultivars, and would include those that have been shown in trials to be useful and acceptable. It would exclude vegetable cultivars with major faults within a crop group, such as low yield, frost susceptibility and poor morphological characters of the part of plant of economic importance or poor storage quality. There is an increasing emphasis on nutritive values of the end product, especially with a view to improving the nutrition status of subsistence farmers and their dependents. A technical background paper presented to the World Food Summit emphasized the importance of vegetables to all income groups and predicted that demand in developing countries would increase by 3.4% throughout the 1990s. The 'descriptive' list would enumerate the pros and cons of a particular cultivar so that a grower is able to decide which one is best for his particular requirements. Information provided would include suitability for a specific soil type, season, market outlet, pest control and disease resistance. Separate lists can be published for different types of growers such as commercial producers and home gardeners. The concept of cultivars for 'continuity of supply' is often important for home garden and self-sufficiency schemes. 'Descriptive' lists are widely used in countries with well-developed agricultural and horticultural industries and can play a major role in developing countries by drawing attention to suitable seed material that is available.

Recommended cultivar lists

This type of list is relatively short and sets out to advise growers which cultivars are firmly recommended for specific crops and purposes. The authority providing the information (usually government or a public sector agency) does not confine the growers' choice of cultivar to those on the list, but clearly advises on the best material available.

Restrictive cultivar lists

A restrictive cultivar list has the objective of restricting the choice of cultivar to those on the list. Cultivars not on the list are prohibited and are not allowed to be offered for sale or to be grown.

Cultivars should not be restricted without good reason, and it is normal practice for a restrictive list to include more cultivars of each crop species than the recommended list. This allows for variations in availability due to seed yield fluctuations resulting from contingencies beyond the control of the seed producer, such as crop failure.

Responsibility for organizing cultivar trials

The cultivar trials should be organized and conducted by an impartial agency. In practice, it is usually a government department such as a branch of a ministry involved with agriculture-related affairs. In the ideal situation, the work is done by an institute or organization that is government-financed, but separate from all other departments.

Individual countries or groups of countries have their own approach to cultivar evaluation and usually delegate responsibility to organizations in existing infrastructures that include the subject of seed in their remits. The primary points to observe are that the trials should not be conducted by organizations involved in the sale of seed or by plant breeders who have a vested interest in the trial results. However, in the present economic climate of more national governments advocating an increased private sector role in conjunction with a decreased public sector role, there is need for caution in assigning responsibilities of cultivar evaluation and recommendation. Despite this trend, the role of governments in assuring food security in addition to protecting vegetable producers from exploitation remains of paramount importance.

Organization of cultivar trials

Separate trials must be organized for each type of crop, even to the extent of conducting separate trials for specific market outlets. For example, if a trial is needed for the evaluation of outdoor tomato cultivars, it is necessary to have separate experiments for the bush (determinate) and trained (indeterminate) types. But whether or not different production systems are included will be influenced by the production techniques used by growers in the locality of the country where the trials are done. Similarly, season of cultivars is also considered; thus, for example, testing cultivars of bulb onions for their suitability for sowing outside in autumn should be independent from a trial to test cultivars for spring sowing.

Design of trials

The trials are usually based on a randomized block design with three replicates. The results are analysed statistically using an analysis of variance. Methods of analysing trial data have been described and discussed in detail by Silvey (1978).

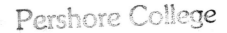

Organizations responsible for trials usually have several sites in the region or country associated with the vegetable crop production areas. In addition to the replication within individual sites, it is usually important to plant (or sow) at several different sites simultaneously. All available sites are not necessarily used for all crops. Most trial organizations have a 'home' ground adjacent to their headquarters and also have substations in key areas. The substations can be on satellite grounds or sites such as university farms and experimental stations of ministries of agriculture.

The AVRDC has an Africa Regional Programme, which conducts trials at Arusha in Tanzania and at other centres in the region.

The use of sites rented from commercial growers is not usually recommended, as it can result in problems relating to access, security or even cultural requirements of the crop.

The trials are normally conducted over at least a 3-year period, which allows for seasonal variations. It would be premature to make recommendations from successive trials in a shorter period. In practice, the established trial organizations run each specific crop trial annually. A standard commercial cultivar that is well known to growers is used as the 'control' or 'check'. This has two advantages: first, material familiar to vegetable growers is included and, second, the control cultivar is a useful basis for comparing factors such as time of harvest and yield. The controls can be replaced after several years by other consistently better cultivars when identified from the results of the ongoing trials.

Allowance must be made to examine a range of harvest dates; for example, spring cabbage cultivars may be suitable for early harvest as 'greens', i.e. before hearting, or may be grown on, and harvested as, hearted cabbage. In order to assess suitability for both these possible harvesting stages, a split plot design is used for the experimental layout. Similarly, with a crop such as Brussels sprouts, it may be considered necessary to examine single harvests but in a succession of four harvesting dates. In this case, 80 plants would be sufficient in the cultivar plot of each replicate to give four sub-plots each containing 20 plants, one sub-plot to be destructively harvested at each nominated date. Details of these experimental techniques for cole crops were described and discussed by Chowings (1974), for carrots by Bedford (1975), for celery, lettuce, leeks, onions and sweetcorn by Chowings (1975) and for courgettes (*Cucurbita pepo* L.) by Higgins and Evans (1985).

The points relating to good field experimentation should be observed; these include the use of guard rows, uniform treatment of the site and uniformity of cultural operations. The trials should be planned carefully in order to gain the maximum amount of information because information not recorded at the time of the trial cannot be retrieved later. The techniques used in vegetable performance trials have been described by Holland (1985).

Husbandry techniques used in cultivar trials should normally be in accordance with local commercial practice. Where more than one centre is being used for a trial, it is important that each adheres to local customs, and dates of cultural operations cannot be dictated in advance because weather and soil conditions may differ between the trial centres. However, if irrigation, herbicides, fertilizers and other inputs are being incorporated, trial coordinators should ensure that stages of crop, moisture deficits and rates of application are clearly defined and adhered to as far as is practical. Further information on conducting cultivar trials has been prepared by Roselló and Fernández de Gorostiza (1993).

Cultivar Identification and Classification

The conduct of cultivar trials discussed earlier emphasizes the usefulness of cultivars for particular environments and purposes. During the course of cultivar trials, and also as a result of observations made during testing for DUS, a great deal of information is obtained about individual seed stocks and the potential crops that can be derived from them. This information can be used to group cultivars according to common features and to classify them so that each cultivar can be clearly described and identified. During the course of this work, the existence of synonyms and near synonyms can be clarified. In addition to synonyms, it is possible that homonyms are found (homonyms are different cultivars that have the same name).

Vegetable seed is normally sold by the seed traders and purchased by farmers and growers on the understanding that the material is of named cultivars. In some situations, it is important that the cultivar be verified from a sample taken from a seed lot, rather than growing on some of the material to observe morphological characters of the mature plants. There are obvious economic advantages in being able to verify a cultivar from a seed sample, which include the saving of time, space and labour. There have been very significant advances in cultivar 'fingerprinting', especially with the applications of DNA profiling, biochemical techniques and computers for the analysis of images. Although some of the so-called newer methods have been developed over a long period, their range and application have accelerated in recent years with significant advances, especially for the more important agricultural crop species. The new developments and their possibilities for cultivar identification have been enumerated and discussed by Cooke and Reeves (1998).

Morphological descriptors

The morphological and other characters that assist in the identification of individual cultivars are of particular interest to the seed producer. The knowledge of a cultivar's observable distinguishing characters enables workers to inspect plants for trueness to type as far as possible before they are used for seed production, i.e. before the start of anthesis. However, because some characters (e.g. flower colour) are only visible at certain stages of the plant's development, it is useful to have a record (or cultivar description), which assists in the identification at several different stages, especially those which can be observed before any pollination takes place.

The visual characters, such as leaf shape, hypocotyl and flower colour, used for identification are referred to as discontinuous characters and may be readily determined at the appropriate plant stage. Characters in this category should be easily identified without disturbing the plant, and as far as possible they should not be significantly influenced by the environment. A good example of the application of discontinuous characters was demonstrated by Bowring (1970) in the classification of runner beans (*P. coccineus*). Bowring first used the three observable *discontinuous characters* in the species, i.e. habit (climbing or dwarf), flower colour and seed colour to make three initial groups. In order to provide more detail about these three groups, he examined the cultivars according to their *continuous characters*. In this example of the runner bean, the continuous characters evaluated statistically were pod length, pod depth, pod width, beak length, seed number per pod, 100 dry seed weight and pod weight. Using this method, Bowring was able to classify the seven cultivars included in the cultivar trial.

Important factors to consider when deciding on morphological characters to use in cultivar classification of vegetables include the stability of the characters from one generation to the next; unstable characters are not useful. Another consideration is the extent to which the environment used for testing can be the same each year and at each centre where the plants are grown. Major environmental differences, such as the amount of available water and nutrients, will affect the characters assessed.

Cultivar descriptions are compiled and used to ensure that the characters and identities of individual cultivars are not eroded as a result of being multiplied from one generation to the next.

Growth chamber and greenhouse testing

These methods of cultivar identification are usually used for the evaluation of a seed sample's purity, and are of particular importance for seed of some vegetable cultivars that are entered into certification schemes. Payne (1993a) has described techniques, procedures and conditions that are recognized by ISTA for the conduct of a range of specific crop species. For the crops listed, the characters used for identification are at the seedling and/or young plant stage. The vegetable species for which these techniques can be applied, and listed by Payne, include lettuce and French beans (accelerated growth in up to 3 and 4–6 weeks, respectively). Lettuce seedling characters that may be used include cotyledon shape, leaf morphology, hypocotyl and leaf colour. The bean characters include the number of days to first flower and flower colour. For carrot, growth at 10–12 weeks, root colour and shape, and also leaf morphology are the determining characteristics. Onion seedling characters that can be determined include leaf base colour from emergence to 7 days; plant height, bulb formation and colour at 4–6 weeks; also bulb shape, leaf attitude and waxiness at 12 weeks.

Laboratory testing for determining cultivar susceptibility to specific pathogens

Verification of a cultivar's resistance (or susceptibility) to specific pathogens is often of prime importance when assessing material for trueness to type. ISTA has assisted the clarification of approved techniques by the publication of a handbook on this important topic (Schoen, 1993). Standards for testing conditions, inocula and media are described in the publication. Hosts and diseases of the vegetable crop species that are detailed include cucumber anthracnose, cucumber scab, tomato leaf spot, bean anthracnose and southern leaf spot or blight of sweetcorn.

Chemical methods for classification of cultivars

There are at least five so-called rapid chemical identification techniques described by Payne (1993b). These include phenol tests, fluorescence tests, use of chemical bases, chromosome number and other chemical testing methods. However, the majority of these are only indicated for the identification of cultivars of agricultural crop species. The fluorescence test may be used to distinguish fodder peas (*Pisum sativum* var. *arvense*) from edible peas (*Pisum sativum* var. *sativum*). When pea seeds are observed under an ultraviolet lamp (wave length of approximately 360 nm), the testas of fodder peas fluoresce but those of the edible peas do not. Cooke *et al.* (1985) have described a method using vanillin to test freshly harvested seed of broad beans (*Vicia faba*) for the presence of cultivars with tannins. However, as the authors point out, vanillin response in the seed and flower colour when it is grown on are complementary data, as the testa is derived from maternal tissue whereas the flower colour observed from growing on could indicate contamination during flowering resulting from unsatisfactory isolation.

The determination of chromosome number can be used as a 'rapid' test to classify beet seedlings (*B. vulgaris*) according to their ploidy, as tetraploids (28 chromosomes), triploids (27 chromosomes) and diploids (18 chromosomes). In the procedure, which is described by Payne (1993b), young roots not exceeding 2 cm long are taken from 14-day-old seedlings, fixed and stained in preparation for microscopic examination.

Care must be taken when using the tests as some of the reagents carry health and safety risks (Payne, 1993b). It is essential that workers using the fluorescence test wear eyeglasses specified for protection from ultraviolet light.

Electrophoresis for testing of cultivars

The biochemical technique of electrophoresis is generally well established and its application for cultivar identification using seed or seedlings is recognized (Cooke, 1992). Reviews of the uses of this technique for crop taxonomic purposes have been made by Smith and Smith (1992) and Cooke (1995a).

Electrophoresis is defined by Cooke (1992) as a technique used to separate charged particles under the influence of an electric field. Different types of support media have been used to hold the charged particles during separation, including polyacrylamide (hence PAGE being an abbreviation of polyacrylamide gel electrophoresis). Cooke has listed four principal and distinct kinds of electrophoresis, which should be recognized in cultivar testing, i.e. Native PAGE (or starch GE), PAGE in the presence of sodium dodecyl sulfate (SDS-PAGE), isoelectric focusing (IEF) and two-dimensional (2D) methods. The techniques used for cultivar identification have been described by Cooke (1992) and Lookhart and Wrigley (1995).

DNA profiling techniques for cultivar identification

The use of DNA profiling techniques and the different approaches of its applications for cultivar identification have been reviewed by Cooke and Reeves (1998). They stated that the profiling techniques can be essentially divided into two technological types, i.e. probe-based and amplification.

The probe-based technology includes restriction fragment length polymorphisms (RFLPs). As most of the more reliable RFLPs depend on radioactively labelled probes and autoradiography, the application of the technology will be restrictive. An example of the use of RFLPs is provided by Smith and Smith (1991), who reported on the 'fingerprinting' of maize hybrids and inbred lines.

The application of amplification-based DNA profiling is relatively new for the identification of cultivars. The technique referred to as random amplified polymorphic DNA (RAPD) is widely used in laboratories specializing in the development of systems for cultivar identification; it is very likely to become important in the topic of cultivar verification as it does not involve radioactivity, is relatively quick and can be successfully automated. These techniques have been described by Caetano-Anolles (1996); an example of their application has been the identification of soybean cultivars as reported by Jianhua et al. (1996).

While the applications of RFLPs and RAPDs have become significant in the development of methods for distinguishing between cultivars, their limitations have been recognized (Cooke and Reeves, 1998). Thus, the so-called second-generation techniques have been attracting attention for the further development of systems for cultivar identification. These more advanced methods include the use of a sequence-tagged site microsatellite system (STMS). Morell et al. (1995) have described these DNA profiling techniques for a wide range of ornamental and crop species including soybean and tomato. Another promising advanced method is the use of the technique known as analysis of amplified fragment length polymorphisms (AFLP), and is described by Vos et al. (1995). This method has been used by Tohme et al. (1996) for the analysis of gene pools in a wild bean collection.

Image analysis for cultivar identification

The various forms of image analysis provide promising techniques for cultivar identification and classification. These include the use of image analysis (IA) recorded by machine, robot or computer, and taken from a range of records such as negative or positive photographic material in addition to live plant material without its destruction. The principles involved in IA have been described by Cooke (1995b,c). Draper and Keefe (1989) have demonstrated the potential for the use of machine vision for the classification of onion bulb shapes. Image analysis has also been used to measure the size and shapes of French bean pods (van de Vooren and van der Heijden (1993), while Sapirstein (1995) has described the identification of some grain crop cultivars using digital analysis. The use of *Chrysanthemum* sp. leaf images has been investigated as an aid to DUS testing (Warren, 1997). Fitzgerald *et al.* (1997) have reported the possible use of image analysis for detecting sib proportion in *Brassica* cultivars. The further development and applications of IA are likely to remove any possible risks arising from the human factor during production of data for storage, retrieval and transmission.

Seed Quality Control

When a vegetable grower has decided on the most appropriate cultivar to use for a specific purpose, it is desirable that he or she be supplied with genuine seed or plant material of that cultivar. A range of schemes that certify the authenticity of the seed sold have evolved in different parts of the world, each of which provides some form or level of assurance that the seed supplied is actually of the cultivar that it is claimed to be. In theory, this ensures verification of the seed stock without the farmer or grower having to see the resulting crop derived from it. However, in practice, the quality assurance depends on both the rigour of the system as well as the discipline and efficiency of the different parties responsible for implementing it. Detailed information on the roles and responsibilities of both the public and private sectors in seed quality control are clearly discussed by Kelly (1994).

Seed Certification

Certification schemes are generally organized on a national basis and a wide range of schemes providing verification of sowing and planting materials have been operating in some countries since the 1920s. It was not until the 1950s that there were steps towards coordinated efforts between countries. These were instigated by the OECD and have become the basis on which most seed certification schemes have been modelled (OECD, 1996). Bowring (1998) has given detailed descriptions of the OECD Seed Schemes.

The Association of Official Seed Certifying Agencies (AOSCA) is the main organization for establishing standards for genetic purity, cultivar identity and standardization of seed certification in North America. This organization also includes some South American countries and New Zealand in addition to Canada and the USA. The Association publishes a handbook on seed certification (AOSCA, 1971). The history, the legal framework, technical achievements and possible future developments of AOSCA were reviewed by Bradnock (1998a). There is close liaison between AOSCA and the OECD Seed Schemes.

The primary aim of seed certification is to check the crop from which the seed is produced and link this verification with agreed minimum standards of other important features of the seed lot. These include health, potential germination and mechanical purity. Thus, the structure of a properly functioning seed certification scheme is complex. This is necessary in order to ensure that all the facets of seed quality are accounted for. Despite the apparent complexity, each part of a seed certification scheme can be clearly defined or described.

Components of a seed certification scheme

The different requirements and standards that are examined separately but which contribute to the composite assessment of a seed lot for certification vary from country to country, but generally the following components or principles are included:

1. A National Designated Authority (NDA), which is appointed by the national government, is responsible for implementing the rules of the scheme on behalf of the government. The NDA acts as the inspectorate of the different parts of the scheme and coordinates the findings of the required tests and observations.
2. Cultivars are accepted into the scheme only when shown to be DUS and have also been shown to be of significant agronomic value.
3. The breeder, or the institute that bred the original cultivar, is responsible for the maintenance of the cultivar and supply of breeder's stock for further multiplication.
4. Each generation of seed is clearly defined, i.e.:
  - Pre-basic: this is seed material at any generation between the parental material and basic seed.
  - Basic seed: this is seed that has been produced by, or is under the responsibility of, the breeder and is intended for the production of certified seed. It is called 'basic seed' because it is the basis for certified seed and its production is the last stage that the breeder would normally be expected to supervise closely.
  - Certified seed: this is the first generation of multiplication of basic seed and is intended for the production of vegetables as distinct from a further seed generation. In some agricultural crops, there may be more than one generation between basic and certified seed, in which case the number of generations of multiplication after basic seed is stated, e.g. first or second.
  - Generally, many of the vegetable crop species have higher multiplication rates than the agricultural crops such as cereals. Vegetable species with relatively low multiplication rates include members of *Leguminosae* such as peas and beans. It is usually the multiplication rate of a species that determines whether or not further generations can be produced beyond the first multiplication from basic seed.
  - Standard seed: this category contains seed that is declared by the supplier to be true to cultivar and purity, but is outside the certification scheme.

5. The agronomic requirements to be observed in planning and production of the seed are defined for specific crops. These include points such as administrative checks on the recent cropping history of the site, its distance from other specified crops and the number of years since previous crops of the same or related species were grown on the site.

6. Tests:

- Laboratory and control plot tests of the stock seed used for production of certified seed.
- Inspection of the seed crop production field and observations by the designated authority. These include trueness to type, isolation and freedom from specific weeds and seed-borne diseases.
- Sampling techniques to be employed.
- Tests to be done in the laboratory and on field plots with samples to check the identity and purity of the cultivar.
- Tests to be done in the laboratory to determine germination, purity and presence of specific seed-borne pathogens.

The success of a seed certification scheme depends on the collection and compilation of evidence from several aspects of seed production and quality control. Furthermore, it also relies on demands for certified seed from vegetable growers. Thus, a successful certification scheme depends on the ability of a country to produce seed of high quality in sufficient quantities to meet market requirements. It is the culmination of successful seed programmes, which over several or even many years have established the required infrastructure of a seed industry. If seed certification schemes in a country aim higher than the industry's current realistic capability, then not only will they fail but their shortcomings or failure will also be detrimental to any future development of seed legislation or seed control in that country.

The Quality Declared Seed System

The Quality Declared Seed System (QDS) was introduced by FAO (1993). Following its initial years of use the organization held an expert consultation to review the ongoing system in order to revise, update and increase the number of crops included.

The revised edition was subsequently published (FAO, 2006). The scheme is of particular value in countries where there are insufficient resources or lack of infrastructure for the establishment of highly evolved seed monitoring systems, such as seed certification.

In recent years the QDS system has been important and of value for the following reasons:

1. Seed purchased for emergency relief supplies.
2. A reference scheme for these purposes, especially in international seed movement.
3. The scheme can also be applied to assist other potential seed suppliers, e.g. farmers groups and cooperatives, private farms and NGOs who enter into seed supply activities for seed quality assurance.

The system has been designed, and further modified, to ensure that a country's existing seed quality control resources are used to maximum advantage. The concept of the QDS system is to enable farmers and growers to have access to seed material of a satisfactory standard. The QDS scheme also recognizes the role of extension services in the demonstration of improved seed to the farming community. QDS is not intended to be an alternative to, or in competition with, more developed seed quality control systems, or duplicate the work of specialist organizations. The documentation of the system refers to varieties rather than cultivars. The scheme recognizes three types of varieties:

1. Varieties developed through conventional breeding technologies.
2. Local varieties that have evolved over a period of time under the particular agro-ecological conditions of a defined area. A local variety is sometimes called a 'landrace' or an 'ecotype'.
3. Varieties developed through alternative plant breeding approaches such as participatory plant breeding.

The required organizational framework for the production of QDS on a government's territory should have the following official organizations established along with appropriate staffing and equipment. These include:

- A seed consultative committee and variety registration system.
- A seed quality control organization.

A QDS Declaration should be made available to the quality control organization for each seed lot (one after planting the intended seed crop and the second following seed processing).

The legal framework of the system requires that a participating country establishes a list of cultivars eligible for inclusion, and that participating seed producers are registered with the relevant national seed authority. The national authority is responsible for checking at least 10% of seed crops entered in the scheme and at least 10% of 'Quality Declared Seed' offered for sale in the country. *The Quality Declared Seed Technical Guidelines for standards and procedures* (FAO, 2006) provides crop-specific sections for some 82 crops, including eight food legumes and 35 vegetables; there is provision for adding further crop species. The crop-specific sections outline the requirements and obligations for each seed crop including facilities and equipment, land requirements, field standards, field inspections and seed quality standards.

'Truth in Labelling'

This type of seed quality control does not have minimum standards but relies on the vendor making a statement as to the quality of seed offered for sale. The seed seller is obliged by law to state certain facts regarding the quality. The statements made by the seller for individual seed lots are subject to random checking by the government seed control agency; most truth in labelling schemes are based on a 10% sampling. The required standards can vary from country to country, and indeed from crop to crop. The truth in labelling concept is a useful way of commencing seed quality control in the early stages of a developing seed industry.

Summary of Tests and Cultivar Trials

- Tests for distinctness, uniformity and stability (DUS); specific for cultivar registration. (UPOV Test Guidelines for a range of vegetables are listed in Appendix 2.)
- Cultivar trials (sometimes referred to as 'variety trials').
- Different kinds of trials assist: identification of synonyms and homonyms.
- Screening prior to breeding programme.
- Grow-out tests (GOT) to check morphological characters of a seed lot.
- Assessment of the suitability of cultivars for recommendation to growers through descriptive lists, recommended lists and restrictive lists.
- Seed organizations may use trials for 'growing-out' tests or for looking at competitors' cultivars.
- The observation of discontinuous and continuous characters in formulating cultivar descriptions.
- Special conditions and requirements for cultivar identification. These include growth chamber and greenhouse testing to observe and determine characters in very young plants. Laboratory tests to determine cultivar susceptibility to specific pathogens.
- Specialist tests include rapid chemical identification techniques, electrophoresis, DNA profiling and image analysis.
- Seed quality control: OECD coordination of seed certification. However, there can be differences between different national schemes. There are clear definable components of a seed certification scheme.
- Definitions of the seed generations of certified seed, i.e. *pre-basic, basic* and *certified*. 'Standard seed' implies true to cultivar and purity, but NOT used as a certification definition.
- FAO Quality Declared Seed Scheme.
- Truth in labelling.

Further Reading

Plant nomenclature

Brickell, C.D., Baum, B.R., Hetterscheid, W.L.A., Leslie, A.C., McNeil, J.C., Trehane, P., Vrugtman, F. and Wiersema, J.H. (eds) (2004) *International Code of Nomenclature for Cultivated Plants*, 3rd edn. ISHS, Leuven.
Douglas, J.E. (1980) *Seed Policy and Development*. Westview Press, Boulder, Colorado, USA.
FAO (2001) *Seed Policy and Programmes for the Central and Eastern European Countries, Commonwealth of Independent States and Other Countries in Transition*. FAO Plant Production and Protection Paper 168. FAO, Rome.

Kelly, A.F. (1989) *Seed Planning and Policy for Agricultural Production.* Belhaven Press, London, 182 pp.

Tripp, R. (2001) *Seed Provision and Agricultural Development.* James Curry Publishers, Oxford.

General

Le Buanec, B. (2007) Evolution of the Seed Industry During the Past Three Decades. *Seed Testing International*, No. 134, October 2007, pp. 6–10.

Spedding, C.R.W. (ed.) (1979) *Vegetable Productivity.* Symposia of the Institute of Biology, No. 25, Macmillan, London, 268 pp.

2 Principles of Seed Production

Before seeds can be produced from a vegetable species, it is necessary for the crop to flower. Plant physiologists have made detailed studies of the flowering requirement in several species, and the reader is referred to texts such as Waring and Phillips (1981) and Atherton (1987).

Some plant species pass from the vegetative phase to the reproductive phase with no special requirement or stimulus, whereas in others there is a clearly defined transition between the two phases. The initial phase before the plant is receptive to the external flowering stimulus is referred to as the 'juvenile' phase. The attainment of the required physiological stage, 'age' or 'puberty' is related to factors such as stage of growth, i.e. the number of leaves rather than actual age as described, for example, by the number of days from sowing.

Species that have a special physiological requirement to pass from the vegetative stage to puberty are generally either dependent on day length (photoperiod) or have a low temperature requirement (vernalization). The induction of flowering in vegetables has been reviewed by Roberts *et al.* (1997) and the timing and prediction of flowering in a range of crops has been discussed by Summerfield *et al.* (1997).

Pollination and Fertilization

The morphology of different types of flowers, pollen formation and the process of fertilization has been documented by several authors, and the reader is referred to Copeland and McDonald (1995). Detailed accounts of insect pollination of a wide range of crops have been described by Free (1970). The theory of incompatibility systems and their practical applications are discussed by Lewis (1979).

The transfer of pollen to the stigma is achieved in the flowering plants by wind, animals or water. In practice, as far as a vegetable seed production is concerned, it is only the transfer by wind and animals, a phenomenon that is normally of concern, and the most important pollinating agents in the animal kingdom are the insects. Flowers that are wind-pollinated are said to be anemophilous, e.g. sweetcorn (*Zea mays* L.) and European spinach (*Spinacia oleracea* L.). Flowers that are insect-pollinated are entomophilous and examples include most of the cole crops (*Brassica oleracea* L.), carrot (*Daucus carota* L.) and onion (*Allium cepa* L.) Some of the species that may on occasions be cross-pollinated by insects are in practice largely, if not completely, self-pollinated. This occurs as a result of flower morphology that is highly adapted to self-pollination; examples include the garden pea (*Pisum sativum* L.), dwarf bean (*Phaseolus vulgaris* L.), lettuce (*Lactuca sativa* L.) and tomato (*Lycopersicon lycopersicon* (L.) Karsten). The modes of pollination for individual crop species dealt with in this book are given in Chapters 7–16.

The Use of Insects to Increase Seed Yield

A very large number of insect species are directly involved in pollination. For example, Bohart and Nye (1960) identified 334 insect species visiting carrot flowers, and Hawthorn *et al.* (1960) identified 267 insect species on onion flowers in a crop grown for seed production. The two most important insect orders concerned with the pollination of vegetable seed crops are the *Hymenoptera* that includes the ants, bees and wasps; and the *Diptera*, the large order that includes the flies. The level of pollinating activity by insects in an entomophilous species grown for a seed crop will have a direct effect on seed yield. In many instances, the seed produced relies entirely on natural insect pollinations in addition to the roving honeybees maintained by bee-keepers. Bohart *et al.* (1970) have reported that in the USA large areas of lucerne (*Medicago sativa* L.) seed production in Idaho and Oregon have 'drained off' populations of pollinating insects. On several occasions, in Sudan, the author has observed virtually no insect activity on fields of onion in full flower; this paucity of insects is attributed to the very frequent applications of insecticides to the cotton crops in the Gezira area. Many specialized vegetable seed producers, especially in the USA, ensure adequate pollinator activity while their crops are in flower by arranging for colonies of honeybees to be supplied on a contract basis (see Fig. 2.1). The same concept applies where seed production is on a smaller scale, but is equally important. Figure. 2.2 illustrates a private beekeeper and his hive in Nepal.

Pollinating insect activity can be increased by improving the microclimate. Drifts of wind-pollinated crops such as maize or sweetcorn are planted by some seed producers to improve the micro-environment. Crops such as sunflowers (*Helianthus annuus* L.) may be used, especially if their flowering does not coincide with that of the vegetable seed crop.

A wide range of pesticides is used in modern farming systems (except in crops produced organically) and many of them are toxic to bees and other beneficial pollinating insects. The protection of honeybees from pesticides has been reviewed by Atkins *et al.* (1977). Generally, cooperation between the

Fig. 2.1. Colonies of honeybees supplied by contractors to seed producers in California, USA.

Fig. 2.2. A Nepalese bee-keeper providing a hive for brassica pollination.

bee-keepers and the pest control industry will ensure that only the safest recommended materials are promoted for the specific purpose of the safety of pollinating insects. Supplementary hives should be located on the perimeters of flowering crops, and placing them in position 2–3 days after pesticide applications will minimize the danger. Pesticides toxic to bees should not be applied while bees or other important pollinators are foraging, and drift from spraying operations on to colonies or flowering crops should be avoided.

Current Problems for Hive Bees and Wild Bees

One of the major issues relating to pollination in recent years has been the decline in the security of hive bees and the reduction in bumblebee populations. Bee-keepers and biologists have outlined the problems that have been identified with these very important and essential pollinating insects. Attempts have been made to bring the matter to the attention of international organizations and national governments, especially in terms of increased funding for related research and extension.

Hive bee problems

Apiarists have always had to be aware of biotic threats to their bee populations. In recent years, new problems have arisen that currently remain unanswered. The most important of these is usually referred to as 'colony collapse disorder'.

Colony collapse disorder (CCD) is a term given to a hive bee colony from which the population has been radically reduced. The symptoms include absence of adult bees (although the queen is present), a young workforce and absence of dead workers either in or near the hive, implying that the missing bees have died while away from their colony. There are several hypotheses as to the cause or causes which include: the Israel acute paralysis virus, pesticides, electromagnetic radiation (including mobile telephones), unidentified pests or pathogens, current practices by bee-keepers (such as use of antibiotics in colony maintenance) and genetically modified organism (GMOs) crops.

However, it must be stated that none of these suggestions has yet been proven. It must also be pointed out that apparent CCD has occurred in some countries where no GMOs were grown. There is a lot of speculation and serious research is required to identify the cause, or causes, as well as satisfactory controls for CCD.

CCD has been reported in the USA, Canada and some areas in Europe. However, it is not clear whether all of these outbreaks all started from exactly the same cause of colony collapse.

Varroa mite (*Varroa* spp.): *Varroa* mite is a pest that causes deformation of larvae in the hive; this mite also transmits the fungus *Nosema ceranea* between and within bee colonies.

Inference for seed producers

The seed producers rely very heavily on bees for pollination of many vegetable seed crops. The presence of CCD in an area may deter the contract hive owners from taking their colonies into affected areas, or a premium may be required to offset their possible losses. This can also be the case with outbreaks of other serious hive bee disorders. The supply of bee colonies by contractors is often called 'migratory bee-keeping', because the hive owners follow the crops that benefit from supplementary pollinators through the season, which is very widely practiced in the USA.

Bumblebee decline

The decline of bumblebee populations and also of the number of their species has been reported by Goulson *et al.* (2008). The main reduction in numbers has been observed particularly in western Europe and North America. The influence of farming practices on the wild pollinator populations in some of the cucurbits has been discussed by Schuler *et al.* (2005).

The main reasons for the decline are thought to be the development of intensive farming practices correlated with areas of high human population (Goulson *et al.*, 2008). The reduction of wildflower and hay meadows, clover leys and use of herbicides has reduced the availability of essential nectar and pollen that are often the only sources of energy for this group of pollinators. In addition, bumblebees have lost some of their nesting sites following reduction of hedgerows or grassy tussocks in rough grassland, depending on their species (Goulson *et al.*, 2008). In the UK, the Bumblebee Conservation Trust advises farmers and growers on suitable actions to take and also techniques to apply for the enhancement of wild bee populations

Climatic factors influencing pollination efficiency

Other factors that can adversely affect the level of pollination include rainfall. Ogawa (1961) reported that continuous rainfall during onion flowering in Japan reduced the seed set significantly and that the reduction of seed yield was proportional to the duration of continuous rain. Dry periods of up to 6h improved the seed set. Local weather conditions not only affect the pollinating insect activity but may be responsible for a reduced rate of stamen dehiscence.

Pollen Beetles

The pollen beetles (*Meligethes* spp.) are largely confined to affecting seed crops in *Cruciferae*. These small blue-black beetles overwinter in hedgerows, but in warmer weather migrate to the crucifers. This pest is often not noticed until 'blind' stalks are observed, and by that time it is too late for an effective control. Spraying with appropriate insecticides should be done when the plants are in the early bud stage unless significant populations of their natural predatory enemies, including species of *Ichnuemonids* or *Braconids*, are observed to be dealing with the potential problem. When producing hybrid seed, parent lines should be checked for the presence of pollen beetle to ensure a satisfactory seed set.

Pollination in Confined Spaces

There are occasions when plants for seeding are cultivated in, or transferred to, enclosed structures such as cages, polythene tunnels or greenhouses (see Chapter 3). Although the isolation provided by the structure normally eliminates the possibility of insects transporting undesirable pollen from outside, it is necessary to ensure that adequate pollination occurs within the enclosure. In some instances, hives or smaller honeybee colonies are introduced and the pollinating insects work effectively. However, in some instances, such as relatively small cages, bees have a strong tendency to attempt to escape from the enclosure and die in the process. Bohart *et al.* (1970) reported that honeybees operating in cages depressed the potential seed yield in onions. They found evidence of physical injury to stigmatic surfaces and that bees stripped pollen from the stigmas. Faulkner (1971) reported that bees tended selectively to self- or sib-pollinate each of the two inbred parents in Brussels sprout hybrid seed production, in contrast to blowflies that pollinated at random and assured a higher degree of cross-pollination between parent lines. The efficiency of blowflies as pollinators of brassica crops was reported by Faulkner (1962) and they are used as pollinating agents for some seed crops of *Cruciferae, Apiaceae* and *Alliaceae* when grown in enclosed environments (Smith and Jackson, 1976).

Effect of Environment on Pollination Method

It is possible for a change in emphasis of a species' pollination method to occur when it is grown outside its natural habitat. This can result in a degree of instability that is not always appreciated. For example, the tomato flower is normally accepted as being completely, or at least predominantly, self-pollinated, but work done in California by Rick (1950) demonstrated that there can be a significant amount of cross-pollination in some locations when specific pollen vector insects are present. The garden pea (*Pisum sativum* L.) is normally accepted as being completely self-pollinated, but Harland (1948) showed that when grown in Peru (an area outside its normal habitat) there was up to 3.3% cross-pollination. These examples demonstrate that the predominant pollination method may modify with provenance of seed production.

Summary of Flowering and Pollination by Insects

- Physiological age of species for flower initiation.
- Wind pollen transfer and insect pollen transfer in vegetable seed crops; (anemophilous and entomophilous pollination).
- Pollinating insects are essential for the satisfactory fertilization of many vegetable crop species (except wind-pollinated crops, such as *Beta vulgaris* types and sweetcorn).
- Increased number of pollinating insects and their activities increases seed yield.

Increase pollinating-insect activity by:

- Choice of seed production area and site.
- Improving microclimate. (See also in Chapter 3, section Modification of the Environment.)
- Short-term or long term-wind protection barriers. (See also Chapter 3, section Modification of the Environment.)
- Other additional crops to attract pollinators.
- Provision has to be made for introduction of either hive bees or blowflies in isolation such as cages and greenhouse structures.
- Use of bees from colonies hired from bee-keepers, or a supply of blowflies depending on crop species and environment.
- Change of location of a crop may mean a change of emphasis of self- or cross-pollination.

Insect-pollinating activities are decreased by:

- Misuse of pesticides.
- Careless timing of applications of pesticides.
- Decline of bumble bees due to loss of nesting sites and pesticides.
- Hive bee problems, including CCD (awareness of ongoing problems and investigations relating to hive bees and bumblebees).
- Other pests, pathogens or hazards of hive bees.
- Pollen beetle: close observation should be made to monitor build-up of pollen beetles prior to start of anthesis, this is especially important in cruciferous seed crops.

Plant Breeding in Relation to Vegetable Seed Production

Hybrid cultivars

There are eight types of hybrid cultivars that have been described by Pickett (1998). The most commonly used types used for vegetables are F1, double-cross, triple-cross and synthetic hybrids. It is vital that the advice and instructions of the plant breeder responsible for maintaining individual cultivars be adhered to during seed production of the various types of hybrids.

F1 hybrids
This type of hybrid is now available in a wide range of vegetable species; it is by far the most widely used hybrid technique used by vegetable breeders. An F1 hybrid is produced by crossing two distinct lines. In practice each of the two parent lines are the result of inbreeding. As far as commercial seed production is concerned, F1 hybrid cultivars are the result of crossing two inbred lines that have been maintained under the control or supervision of plant breeders, and which are known to produce a desirable hybrid. Since the late 1940s or so, we have seen the increasing development and use of F1 hybrid vegetable cultivars. Riggs (1987, 1988) has reviewed the breeding of F1 vegetable cultivars. Choudhury (1966) discussed the exploitation of F1 vegetable hybrids developed for India and Takahashi (1987) has described the utilization of hybrid vegetable cultivars in Japan.

The advantages of F1 hybrid cultivars include uniformity, increased vigour, earliness, higher yield and resistance to specific pests and pathogens, although these factors are not always all present in any one cultivar of a vegetable species. Heat tolerance has been an important character in the development of F1 Chinese cabbage for tropical areas.

Theoretically, all the plants in an F1 hybrid cultivar resemble each other exactly, but because some self-pollination of the female parent used in the cross may take place, some plants that are not F1 hybrids may occur and are usually morphologically different. These off-types in an F1 hybrid are usually the result of accidental self-pollination of the female parent and are generally known as 'sibs'. These are so-called because they are the result of 'sister' or 'brother' plants crossing or selfing with the female line. The incidence of sibs in seed production of Brussels sprouts and the influence of pollinating insects on the percentage of a seed lot have been investigated by Faulkner (1978), who demonstrated that blowflies pollinate in a more random manner than honeybees; the flies drastically reduced the percentage of sibs under glass and polythene structures and to a lesser extent in cages.

In addition to the problems associated with sibs in an F1 hybrid seed lot, there are increased production costs compared with open-pollinated cultivars due to the following factors. The development of the initial breeding programme; subsequent maintenance of the inbred parents; extra land required to allow for male parents; care with sowing, isolation and harvesting; high labour input when flowers of the female parent have to be emasculated (especially when hand emasculation is necessary) and the lower seed yield per unit of land sometimes experienced with production of F1 hybrid cultivars.

BREEDING SYSTEMS UTILIZED IN F1 VEGETABLE CULTIVAR PRODUCTION   The different systems utilized in the seed production of F1 vegetable cultivars have been reviewed by Wills and North (1978); details of the genetic principles involved are described in leading plant-breeding texts such as Frankel and Galun (1977). The following are the main methods applied for the production of F1 hybrids with crop examples to which the principle is applied:

- Self-incompatibility (SI), e.g. Brussels sprouts. This results from the presence of the so-called S-allele that prevents self-fertilization.
- Cytoplasmic male sterility (CMS), e.g. carrot, onion, parsnip. This type of male sterility is controlled by the cytoplasm and has been described by Innes (1983).
- Gynoecious lines, e.g. cucumber. The gynoecious parent has only (or predominantly) female flowers, thus crossing only takes place with the male parent.
- Dioecy, e.g. spinach (Spinacea oleracea), use one line of male and another of female, but rogue out any plants showing male flower characteristics in the designated female line.
- Monoecy, e.g. sweetcorn. In male and female inflorescences borne on different parts of the same plant, the male inflorescence is either manually removed or chemically suppressed before anthesis.

- Manual control, e.g. tomato and aubergine. It involves emasculation of individual flowers on the female parent before dehiscence of the stamens, followed by manual addition of pollen from the designated male parent.
- Chemical suppressants, e.g. some cucurbits such as squash (including courgette and vegetable marrow). A chemical suppressant, such as ethephon, is applied to stimulate the production of pistillate (male) flowers.
- Chemical hybridizing agent (CHA) involves application of a chemical to cause pollen abortion that results in male sterility. It is essential that the chemical does not result in ovule abortion. While this technique has been adopted for some agricultural crop species, including cereals and maize, it has so far only been of research and development application for vegetable crop hybrids.

SYNCHRONIZATION OF FLOWERING OF MALE AND FEMALE LINES   Where the pollen donor (male) and pollen receptor (female) parts are on separate plants as in the production of hybrid seeds, it is essential that the shedding of pollen and receptiveness of the stigma occur simultaneously. Seed producers refer to this synchronization of flowering as 'nicking'. The matching of parent lines for hybrid seed production therefore depends upon their ability to 'nick', in addition to the agronomic potential of the progeny. Parents of hybrids that 'nick' in one location do not necessarily flower at the same time or duration at another. Nicking may also vary from one season to another. Synchronization of flowering between parent lines of some vegetable species can sometimes be assured by sowing time of the pollen parent in relation to the female as well as successive sowings of the male parent in order to ensure a satisfactory overlap of anthesis. Information of these requirements should be provided by the plant breeder supplying the inbred parents, but it will also be learned from experience in a specific location.

PHILOSOPHY AND POLICIES RELATING TO PRODUCTION AND USE OF F1 HYBRID CULTIVARS   Hybrid seed is several times more expensive than open-pollinated seed for a given crop. This is because the hybrid seed producer has to recover the investment incurred for developing it and pass on to the consumer the increased costs of hybrid seed production. In some crops the potential F1 seed yield is lower per unit area than for open-pollinated crops of the same species. One main advantage of an F1 hybrid to the organization developing and marketing it is the relative difficulty with which competitors can reproduce the cultivar; further protection is afforded with the introduction of plant breeders' rights. Regardless of legal constraints, a grower cannot successfully use seed saved from F1 hybrids due to segregation in the following generation.

It is sometimes suggested that F1 hybrids provide commercial seed companies with a hold on the market because once growers have found an F1 cultivar to be acceptable, they will be obliged to continue using it in subsequent years. However, it is the author's observation that commercial vegetable producers worldwide are very keen to adopt F1s that have been demonstrated to be beneficial provided that they can afford the extra seed cost. This assumes that there are no other constraints to maximize the cultivar's potential. None the less, it must be emphasized to subsistence farmers that hybrid cultivars' uniformity of maturity in crops from which there is only one harvest per plant (such as cabbage) may not provide the continuity of supply that they expect from some open-pollinated cultivars. There are some schools of thought among organic vegetable producers that resist the use of F1 hybrid cultivars.

It is stressed that individual hybrid cultivars should be judged on their agro-nomic and economic advantages to the grower.

Double-cross hybrids

The double-cross hybrids are produced by crossing two pairs of inbred lines. The two resulting F1s are then crossed to produce a double hybrid. This tech-nique has been used in the production of some *Brassica* and sweetcorn culti-vars, and is considered better than single-cross hybrids because the seed yield is generally higher, although the maintenance of the appropriate lines and crosses increases the isolation requirements.

Triple-cross hybrids

This type of hybrid was described by Thompson (1964) as a method of over-coming the relatively high costs of maintaining inbred lines. Although not of major significance in vegetable crops, it has become important in the produc-tion of some fodder brassicas.

Synthetic hybrids

A synthetic hybrid is produced by the mass pollination of several inbred lines selected for their satisfactory combination ability. Random cross-pollination occurs between the different inbred lines, resulting in a mixture of hybrids. The individual inbred lines to be used for the production of these synthetic hybrids are determined by the breeder. Because of the random crossing that takes place, there may be some variation from one season to another when the same parents are used. Cross-pollination is normally assured because of the individual inbred lines' relatively high levels of self-incompatibility. This system has been used for some cultivars of *Brassica* species, including sum-mer cabbage.

Genetically Modified Organisms

The development of biotechnology and its introduction into the seed industry have introduced a new tool for cultivar production in addition to traditional, or orthodox, plant breeding methods outlined above. Techniques for the transfer of genetic material between otherwise incompatible species or genera have become possible. The techniques used are known as 'genetic engineering'. The resulting cultivars arising from application of the technology are referred to as 'genetically modified organisms' (GMOs); cultivars of this origin are also referred as 'transgenics'.

The methods used to produce genetically modified plants have been described and discussed by Cooper and MacLeod (1998) who also outline the regulation of genetically crops with particular reference to the EC and the UK. They also outline the regulation in the USA, Canada and Japan.

Possible applications of genetic engineering include quality improvement of pest and pathogen resistance, quality, resistance to herbicides, producing male sterility, extension of postharvest storage potential (including shelf life) and nutritional values of a cultivar.

Much of the earlier work relating to crops was led by international companies, but some countries have released GMO material for multiplication by farmers. The development and potential of GMOs has become a very contentious issue. The reader is referred to the relevant publications in the Further Reading list at the end of this chapter including the *Cartagena Protocol on Biosafety to the Convention on Biological Diversity*.

The possible sources of contamination of orthodox (non-GMO) seed crops are outlined on page 60; the role of ISTA in the evaluation of seed lots for GM materials is given on page 107.

Importance of Available Water During Anthesis and Seed Development

Most experimental investigations into the water requirements of crops have been directed at the market yield of the vegetative stage. Very little work has been done into the effects of soil moisture deficit on seed yield and quality. However, there are many indications from the literature that can be extrapolated by the seed producer.

Salter and Goode (1967) reviewed the available information on different crop groups. Generally they found that most annual crops have moisture-sensitive stages from flower initiation, during anthesis, and in some species during fruit and seed development. Legume crops have received the most attention as to the effects of soil moisture shortages on seed yield. The literature indicates that provided there is sufficient water available prior to flowering for plant growth to proceed without permanent wilting, there is little influence on seed yield. But the plants are very sensitive to moisture stress during anthesis when adequate soil moisture, supplemented where necessary by irrigation, will generally provide maximum yield increases. Salter and Goode found that some investigators working with legumes reported an increase in the number of seeds per pod as a result of additional irrigation applied early in anthesis, and that further irrigation during pod growth increased the 1000 grain weight. This general effect in leguminous vegetables is believed to be due to a reduction in root activity when the plants are flowering. The evidence for this on annual leafy vegetables is not clear, especially as most experimental work has been directed at increasing leaf or bud yield rather than seed. The fresh market yield of leaf vegetables such as lettuce and cabbage is generally in proportion to the amount of water received. Some workers have considered that this vegetative response is a precursor to flower formation.

The biennial vegetable crops comprise a wide range of morphological types at the time of marketing, including root, bulb, leaf, petiole or reproductive tissue (e.g. the 'curd' of cauliflower). These different crop types usually display a response to irrigation during growth and development in the first year by an increase in size of their storage organs. There is also evidence that seed yields of these biennials increase with the available water during anthesis. The work on the 'fruit' vegetables (e.g. *Solanaceae*) indicates that they are responsive to adequate available soil water once fruit setting commences. Salter (1958) reported that water shortage during anthesis and fruiting of tomato reduced fruit set and size. This pattern was generally true for satisfactory vegetative development for all indeterminate fruiting vegetables (Salter and Goode, 1967).

Nutrition and Flowering

There is a tendency for there to be antagonism between the mineral nutrient requirements for optimum vegetative growth and optimum reproduction. Wareing and Phillips (1981) reported that low levels of nitrogen tend to result in earlier flowering of some long-day plants. Investigations into cauliflower seed production showed that when nitrogen was applied at 50, 150, 250 and 350 kg/ha, the highest level delayed flowering by up to 10 days (R. Raut, Bath, UK, 1984, personal communication). Liaw (1982) in similar work with seed production of French beans (*P. vulgaris*) found that the higher levels of phosphorus increased flower number. The environmental influences on development, growth and yield of vegetable crops have been reviewed by Krug (1997). The correlative growth of vegetables, including the interactions between vegetative and reproductive parts, and also interactions among reproductive parts have been reviewed by Wien (1997c).

Plant Breeding Techniques, Hybrid Seed and GMOs

- Advantages of F1 hybrids for inclusion of useful and desirable characters in a single cultivar.
- Breeding systems and methods used by plant breeders for production of hybrid cultivars.
- Practical aspects for field production of hybrids, including synchrony of flowering in both parents.
- Possible advantages of GMO cultivars to the vegetable grower and consumer.

Importance of Available Water to the Mother Plant and Mother Plant Nutrition

- Moisture sensitivity during anthesis and other possible stages according to crop.
- Effects of nitrogen and other elements on seed production. (The micronutrient requirements of individual crop species that have a significant effect on yield or seed quality are discussed for the relevant individual crops in Chapters 7–16.)

Further Reading

Seed

Copeland, L.O. and McDonald, M.B. (1995) *Principles of Seed Science and Technology*, 3rd edn. Chapman & Hall, New York.

Flowering and pollination

Atherton, J.G. (ed.) (1987) *The Manipulation of Flowering*. Butterworths, London.

Choudhury, B. (1966) Exploiting hybrid vigour in vegetables. *Indian Horticulture* 10, 56–58.

Delaplane, K.S. and Mayer, D.R. (2000) *Crop Pollination by Bees*. CAB International, Wallingford, UK.

Proctor, M. and Yeo, P. (1973) *The Pollination of Flowers*. Collins, Glasgow.

Roberts, E.H., Summerfield, R.J., Ellis, R.H., Craufurd, P.Q. and Wheeler, T.R. (1997) The induction of flowering. In: Wien, H.C. (ed.) *The Physiology of Vegetable Crops*. CAB International, Wallingford, pp. 69–99.

Wareing, P.F. and Phillips, I.D.J. (1981) *Growth and Differentiation in Plants*, 3rd edn. Pergamon, Oxford and New York.

GMOs

Anonymous (2000) *Cartagena Protocol on Biosafety. The Secretariat of The Convention on Biodiversity*. World Trade Centre, Montreal, Canada.

Goulson, B., Lye, D.G. and Darvill, B. (2008) Decline and conservation of bumble bees. *Annual Review of Entomology* 53, 191–208.

Ferry, N. and Gatehouse, A. (eds) (2008) *Environmental Impact of Genetically Modified Crops*. CAB International, Wallingford, UK.

Thomson, J. (2007) Genetically modified crops – good or bad for Africa? *Biologist* 54(3), August 2007, 129–133.

Tzotzos, G.T. (1995) *Genetically Modified Organisms: A Guide to Biosafety*. CAB International, Wallingford, UK.

3   **Agronomy**

Seed Production Areas

The main seed production areas have become established where climatic factors ensure a relatively suitable environment for the production of satisfactory vegetable-seed crops. These factors include sowing and growing conditions; sufficient rainfall to ensure complete development and maturation of the seeds, but a relatively dry summer and autumn with little rain and wind to enable the seeds to ripen and the harvesting operations to be completed with minimal deterioration of the seed or loss of crop. This is especially important for the dry-harvested vegetable seeds. The lack of inclement weather during the final stages of seed development and ripening is also important from the point of view of disease control on the mother plant as a low relative humidity with minimal rainfall and moderate temperatures minimizes the development and dispersal of many pathogens (Gaunt and Liew, 1981); Garry *et al.* (1998) have also indicated the importance of growth stage and disease intensity in the effects of *Ascochyta* blight on peas. The same concept applies to seed crops extracted from wet fruits such as tomatoes and peppers, especially for those pathogens which affect the mother plant and are also seed-borne.

The suitability of a system for the classification of agroclimatic and soil conditions based on the formulation of the 'FAO Variety Description Forms and Variety Passports' has been described and discussed by Kelly (1994). The system was initially formulated for specific major cereal crop species (i.e. maize, rice, wheat, millet and sorghum). The four main climatic types are designated 'tropics', 'subtropics with summer main rainfall', 'subtropics with winter main rainfall' and 'temperate'; each of these types is subdivided into 'warm', 'cool' and 'cold'. A further subdivision is made depending on annual rainfall; i.e. 'arid', 'semi-arid', 'sub-humid' and 'humid'.

Some of the climatic factors necessary for the biennial seed crops include a relatively mild winter to ensure minimal loss of overwintered plants, although for some species there must be sufficiently low temperatures for vernalization.

The soil's physical properties have also played a part in the development of seed production areas as generally only those soils with a relatively high water-holding capacity are most suited, although they should not be water-logged in winter. Winter soil conditions are especially important when biennial mother plants are lifted for selection and replanted or young plants are trans-planted into their final seeding quarters. The nutrient status of soils can be modified by applications of appropriate macronutrients or micronutrients, but those soils with a satisfactory cation exchange capacity are most useful.

Several classical examples of vegetable seed production areas can be cited; they include parts of North America, especially the Pacific North-West, Northern Italy, Hungary and parts of Australasia. The oceanic effect in coastal areas prevents over-drying of unthreshed material in the field and can assist in reduc-ing loss from shattering prior to harvesting. Excellent farming traditions cou-pled with the capability to apply sufficient quantities of satisfactory quality irrigation water as and when required have also ensured that areas such as California, USA, lead the world's vegetable seed production technology. There are many other highly developed vegetable seed production countries including the Netherlands and France; other areas specialize in specific crop groups, for example, parts of East Africa have been utilized for the production of legume seed crops. The increased market demand for pre-packed small spinach leaves has resulted in the further development of suitable areas in Denmark for the production of spinach seed which is wind-pollinated.

As the agricultural and horticultural industries develop in different countries, so do their requirements for seed and planting materials. Hrabovszky (1982) discussed the possible effects of crop production in developing countries on seed development plans. There are already many areas where there have been substantial developments of vegetable seed production despite adverse climatic conditions; notable examples include Egypt, India, Israel, Kenya and Mexico. This type of development frequently combines climatic attributes such as high summer temperatures with technical inputs such as irrigation. In addition some of the newer areas have often depended on migrant technologists or outside assistance for their emergence as vegetable seed production areas.

The increased market demand for organic seed has led to the exploration of areas where some of the biennial seed crops, such as onion, can be success-ful because of a better environment and prospects for production of disease-free seed where crop protection chemicals are not to be used, although this at first may imply lower yields and technical training.

Modification of the Environment

Within each of the production areas, there are local gradations of climate which result from differences in altitude, topography, proximity to the coast and the presence of natural shelter belts such as woodland. The microclimate in these areas may be further modified in one of the several ways.

The use of shelter and windbreaks

Windbreaks or other forms of shelter are especially useful for vegetable seed production where individual blocks of a seed crop are not always as large as those for agricultural crops.

Solid barriers such as walls divert the windflow, but cause turbulence which can damage the plants. However, windbreaks that offer approximately 50% obstruction provide a relatively extensive shelter with the minimum of wind gusts. A permeable windbreak can decrease the windspeed in the horizontal direction downwind for a distance equivalent to up to 30 times its height, although most shelter is within a horizontal distance on the leeward side of approximately ten times the height of the shelter.

Shelter belts are also beneficial to the individual plants in the crop because water loss by transpiration and evaporation from the soil is reduced. There is less leaf damage by bruising, and the protective waxy layer on the leaf surfaces of species such as peas and onions remains more intact. In coastal areas, windbreaks reduce the incidence of scorch from wind-borne salt originating from sea spray. There is usually a slight increase in soil temperature which is especially useful in temperate areas. Soil erosion from 'blowing' is also reduced.

In addition to the above effects, the improved microclimate will enhance flower development and increase insect activity and pollen transfer in entomophilous crops; the decrease in physical damage to flowers will also result in more fertilization.

Some specialist seed producers believe that the incidence of undesirable cross-pollination between different blocks of cross-compatible crops is reduced when separated by belts of maize or other suitable crops. In practice, ten rows or so of these temporary shelter crops will increase the activity of bee colonies within the separate blocks and so reduce the amount of outcrossing to other blocks. This is especially useful in crops such as aubergine and peppers, where up to 10% outcrossing with other blocks may otherwise be expected. In these instances shelter belts are not a substitute for appropriate isolation distances (see below) but rather enhance them.

Dark (1971), investigating the cross-pollination of sugarbeet, suggested that hedges may affect the incidence of wind-borne pollen in a plot. He proposed that work should be done to see if the sharp drop in pollen concentration on the windward side of a hedge and the gradual build-up again downwind from the hedge were effects of wind deflection or deposition of pollen into the hedge.

There are many different types of windbreaks, but in principle they are either permanent or temporary.

Permanent windbreaks
These are provided by planting lines or belts of single or mixed species of trees that are tolerant of local conditions and have a growth rate suited to quick establishment. A wide range of species is used: evergreens such as conifers and *Eucalyptus* spp. and deciduous species such as *Populus* spp., *Salix* spp. and *Tamarix* spp. Tolerance of the species to salt should be taken into account

where high soil salinity is known to be a potential problem. In many areas appropriate species are planted either along water courses and irrigation channels, or large land areas are subdivided into appropriate size plots by lines of trees.

Temporary shelter

This can be of two types: living materials or manufactured materials. Usually with living plant materials, another crop is grown for its own value and has the dual role of providing shelter for an adjacent crop. Examples depend on the farming system, but *Zea mays* L. (either maize or sweetcorn, as in Fig. 3.1), *Helianthus annuus* L. (sunflower) and the cultivated *Sorghum* spp. are widely used. Occasionally in small-scale operations climbing plants such as *Phaseolus* spp. on a framework provide shelter.

Other forms of temporary shelter can be provided by erecting screens of manufactured materials such as plastic mesh or hessian. Plastic materials should contain ultraviolet (UV) light inhibitors to prolong their useful life by minimizing UV degradation. These types of structure are relatively low in height but have the advantage of immediate effect and they do not compete with crops for water or nutrients.

Possible disadvantages of living windbreaks

The roots of the species used may enter water courses and drains with the result of water loss, blockages or structural damage to pipes and conduits. Dense windbreaks can reduce the available photosynthetic light and will compete for water and nutrients.

Fig. 3.1. Watermelon seed crop with enhanced environment provided by a maize crop.

Living windbreaks can be hosts to a wide range of pests and pathogens; for example, some *Populus* spp. are alternative hosts to the lettuce root aphid (*Pemphigus bursarius* L.), also *Salix* spp. to the willow-carrot aphid (*Cavariella aegopodii* Scop.), which is the vector of carrot motley dwarf virus.

Although the relatively dense windbreaks offer shelter to a range of bird species, there is no experimental evidence to suggest that they lead to an increase in crop loss from bird damage, though this can be possible in some locations.

Cover crops

In some areas of the world vegetable seeds are produced as cover crops in plantations. This is relatively common in the tropics and is partly a reflection of the level of seed industry development and partly because of existing agricultural crop systems, including a strong tradition to provide shade for some crops. Young plantation crops such as palms, citrus and banana offer some shelter and in addition there is sufficient arable space for vegetables in the early life of the plantation.

Preparation, Sowing, Planting and Mother Plant Environment

In general the principles and practices to establish the crop are the same as for the production of vegetables for market outlets. But as the final objective is to obtain seeds to be used for the production of further crop generations, it is important that every care is taken to avoid admixture of seeds or plant material at all stages and produce a typical crop of the cultivar, so that its genetic quality or 'trueness to type' can be fully evaluated.

Crop rotation

Satisfactory intervals between related or similar crops are a standard agronomic practice and the main reasons for crop rotation include plant nutrition, maintenance of soil physical conditions and minimizing the risk of soil-borne pests, pathogens and weeds common to individual crop groups. In addition to these generally accepted reasons for crop rotation, the seed producer has to minimize the risk of plant material or dormant seeds remaining in the soil from previous crops, which are likely to cross-pollinate or result in an admixture with the planned seed crop; thus, information of previous cropping on a given site is an important consideration in seed production. In the case of certified seed production, the minimum number of years free of specified crops is specified by the certification authority. Plants derived from dormant seed or vegetative parts of any previous crops are usually referred to as 'volunteers'. The term 'admixture' usually refers to the addition of seed material that does not conform to the relevant specification for purity of the seed lot in question; it also refers to related genetic material that has been inadvertently mixed in during sowing, harvesting or processing. Therefore, attention must be paid to the number of years since a related crop was grown on the same site and the longevity of seeds of weeds or related crop species compared with the intended seed crop. These periods should be stringently adhered to for the production of pre-basic, basic and certified seed categories and, in practice, for the production of any seed class, whether or not there is relevant or enforced seed legislation.

The need for organic seed producers to use crop species in *Leguminosae* to increase the nitrogen status of their soils must not be allowed to compromise the requirement for sufficient length of time between related crops, which could add to the risk of disease or genetic contamination from dormant seed or volunteer plants in subsequent seed crops. Conversely, the inclusion of legume and/or green manure crops in the rotation can help to increase the time break between crop species with common seed-borne pests and pathogens. However, it should be emphasized that all the seeds used in this rotation in an organic seed production system should also have been produced organically (or in accordance with permitted practice of the prevailing scheme).

Plant population

The main objective of the commercial seed producer should be to achieve, per unit area of land, the maximum yield of seed with the highest possible qualities according to the seed class or category being produced. Therefore, the spacing of the mother plants to produce the intended seed crop can be critical, depending on the species. The potential seed yield per individual plant may be greater than that which is actually achieved per plant in a crop of that species; this may be because the spacing of the plants in the seed crop takes into account the need to reduce the overall duration of successive seeds maturing in the seed crop; this is especially important for seed crops which are harvested in a single pass. Greven *et al*. (1997) working with French beans reported seed yields 2.5 times greater from high-density crops than from a low density. However, in some crops, such as tomatoes, there may be successive harvests at which fruit is harvested in its order of maturity. An interesting effect of increasing the plant densities of some seed crops in *Apiaceae* has been described by Gray *et al*. (1985b); they demonstrated that closer mother plant spacing of parsnip increased the proportion of seed from primary umbels, thus improving seed quality because the embryos of seeds from primary umbels are larger than in seeds derived from subsequent umbels. The same concept has been applied to carrot seed production in which higher densities than those traditionally used have reduced the amount of inflorescence branching and resulted in seed being harvested with more uniform embryo size (Gray and Steckel, 1983); see research and carrot seed production, Chapter 13.

The usual sowing and planting distances within and between rows are given for individual vegetable seed crops in Chapters 7–16. The crop spacings given assume that the mother plant population is growing in a weed-free environment. It should be emphasized that weed control is especially important in seed production. Modifications to recommended plant densities should be considered when an organic seed crop is being produced and a lower plant density will reduce the incidence of some pathogens. In addition to all the accepted reasons for eradicating interspecific competition, it is essential that the quality of the seed to be harvested is upheld. Additional care is necessary to eradicate parasitic weeds such as dodder (*Cuscuta* spp.) and *Orabanche* spp., weeds which are alternative hosts to seed-borne pests and pathogens, weeds which can impede mechanical harvesting and any weed species whose seeds can present difficulties when upgrading the harvested seed lot.

Summary of Seed Production Areas, Shelter, Micro-environment and Preparation

- Main examples of seed production areas, and their climatic characters.
- Classification of agroclimatic and soil conditions, also reference to time required for production of biennials.
- Identification and development of new production areas.
- Modification of the environment: use of windbreaks, theory of windbreak shelter, permanent or temporary shelter, advantages and disadvantages.
- Admixture danger, and production of true to type plants for seed production.
- Extra importance of crop rotation for seed production, risk of volunteers.
- Plant population, spacing and plant densities for weed control and optimum quantity and quality of seed produced.
- Importance of weed and parasitic plant control for production of *satisfactory* seed lot.

The Use of Cages and Protected Structures

Cages

The primary reason for the use of cages is to provide isolation by exclusion of pollinating insects. Having achieved this isolation, it is then necessary to either hand pollinate the flowers or provide appropriate pollinating insects, unless the crop is largely self-fertilizing. The provision of insect pollinators was discussed in Chapter 2. Cages do not exclude stray pollen of wind-pollinated crops (see Fig. 3.2).

Large cages are sometimes used by seed producers for the final quantity of plants from a positive selection for basic seed production. Cages for this purpose may be sufficiently large for up to 100 plants. Smaller cages are widely used by plant breeders for the production of parental material and pre-basic

Fig. 3.2. 'Walk-in' isolation cages fitted with bee colonies.

seed stocks in addition to their activities associated with cultivar development. Some seed producers cover the wooden or metal greenhouse structural framework with netting to maintain an ambient atmosphere which also provides isolation from insect-borne 'foreign' pollen.

Greenhouses

Greenhouses and plastic tunnels are both used for seed production, especially in temperate areas. The former has been used as a convenient working enclosure and for the improvement of plant environment and isolation. Since the advent of large walk-in plastic tunnels, they have been increasingly used for seed production. Plastic clad structures are especially useful in temperate areas as the protected environment for both crop and pollinating insects will produce good yields of high-quality seeds. Plants are usually grown directly in the soil substrate; either sown *in situ* or material transferred from field selections can be replanted in them. Jackson (1985) has discussed the use of polythene tunnels for the production of vegetable seed in the UK and Dowker *et al*. (1985) have outlined the use of polythene tunnels for the production of F1 hybrid onion seed.

These structures are also useful for the production of virus-free seed stocks of crops subject to insect-transmitted viruses provided that the mesh is small enough to exclude the vectors; for example, lettuce seed may be produced in aphid-proof structures to minimize the transmission of lettuce mosaic virus (Fig. 3.3).

The introduction of wild flower strips between tunnels or outside plots is considered to be useful for encouraging the presence of natural insect predators to assist in control of pests; this can be of particular use in organic seed production programmes.

Fig. 3.3. Lettuce seed crop flowering in an insect-proof structure.

The Maintenance of Cultivar Purity

The maintenance of cultivar purity during seed production is achieved by the combination of several factors, some directly relate to agronomy while others result to satisfactory crop planning.

Isolation

One major factor during the course of seed production is to ensure that the possibility of cross-pollination between different cross-compatible plots or fields is minimized. This is achieved either by ensuring that the crops which are likely to cross-pollinate are not flowering at the same time (i.e. isolation by time) or that they are isolated by distance.

In addition to the question of undesirable cross-pollination, adequate isolation also assists in avoiding admixture during harvesting and the transmission of pests and pathogens from alternative host crops.

Tolerance limits to genetic contamination

The maximum permissible or acceptable contamination resulting from undesirable cross-pollination will depend on the species and the class of seed to be produced. It follows that a cross-pollinated crop species will have a higher degree of variation than a self-pollinated crop. While it may be thought desirable to reduce pollen contamination to zero, the amount of permissible contamination will vary with the species and the purpose for which the seed stock is intended (i.e. the intended seed class). Even if it were possible to exclude pollen contamination completely, it would not be possible to have a tolerance limit lower than the species' mutation rate (Bateman, 1946).

The higher the class (or category) of seed, the lower the acceptable number of off-types and these tolerated levels are specified by seed certification authorities or seed quality control agencies according to the crop species and seed class.

Isolation in time

This type of isolation is possible within individual farms or multiplication stations; with careful monitoring, it is also possible in 'seed production villages' where only one cultivar of a crop is cultivated. In the planning stage, it is arranged that the cross-compatible crops are grown in successive years or seasons. This principle is easier to achieve in those areas of the world where the climate allows two successive seed crops to be grown in a year. Seed production centres, or organizations, responsible for the multiplication of relatively few cultivars can plan production so that no two cross-compatible cultivars are multiplied simultaneously.

Despite the possibility of isolation in time, there still remains the need to ensure that cross-compatible crops are isolated by distance.

Isolation by distance

This type of isolation is based on the concept that if a seed crop is sufficiently distant from any other cross-compatible crop, then the adverse pollen contamination will be negligible. In practice, it is impossible to prevent completely 'foreign' pollen reaching a crop because the wind can carry pollen grains or pollinating insects over relatively long distances.

Regulations or recommendations for isolation distances of specific crops take into account the method of pollination (i.e. whether the species is predominantly self- or cross-pollinated) and the pollen vector (i.e. insect or wind). In some countries (e.g. the Netherlands), the minimum isolation distance between different groups or types of cultivars of the same species is greater than for cultivars of a similar type. For example, the minimum isolation distance between runner bean cultivars with different flower colours is greater than for cultivars with the same colour. The specified distances are also greater for classes of seeds to be used for further multiplication than for those distributed to growers for production of a market crop. The generally accepted isolation distances for individual crops are given in Chapters 7–16, but these are only guidelines as the regulations vary between countries. Much of the experimental work investigating optimal isolation distances has been done in temperate regions and factors such as topography, prevailing wind, insect species and insect populations can influence the efficiency of isolation distances.

In practice, there are several possible outside sources of contaminating pollen during anthesis of a seed crop, examples include:

- Compatible crops in cultivar trials.
- Private gardens.
- Workers' crops.
- Cross-compatible commercial crops.
- Volunteer plants.
- Wild or escape species.

Sources of contamination from genetically modified organisms (GMOs)

Other sources of potential contamination from GMOs during the course of vegetable seed production will increase as the number and types of GMOs released in the field increases.

Present examples of these sources of contamination from GMOs may include:

- Admixture with GMO material during harvesting, seed processing or packaging, usually referred to as 'adventitious' material.
- Genetically engineered male lines designed for hybrid production.
- Glyphosate-tolerant cultivars or lines.
- Any cultivar developed for increased shelf life, e.g. tomato cultivars (including those for marketing fresh and those developed for processing).
- Cultivars engineered for changes in starch composition, e.g. maize for agricultural outputs.
- Cross-compatible materials within *Cruciferae* which have been designed for agricultural or biofuel outlets.
- Undesirable characters which have entered or contaminated other seed crops, weeds or garden plants with a common period of anthesis.

Zoning

In addition to the primary isolation requirement for a seed crop there are, in some countries, zoning schemes for specified geographical areas which control the species to be grown either for market or as seed crops in specified areas. In principle, the specification of what is allowed to be grown in a given zone ensures that cross-compatible species, subspecies or types of related crops do not freely cross-pollinate. For example, there are different types of *Beta* species which are all cross-compatible and largely wind-pollinated, and they include mangel, fodder beet, sugarbeet, Swiss chard and red beet. By allowing only one of these types to be grown in a specified area or zone, the chances of highly undesirable cross-pollination between any combinations of them will be greatly reduced. Other horticultural crop groups for which zoning arrangements are made in some countries include *Brassica campestris* (turnip), *Brassica oleracea* (cole crops) and *Allium* species. Zoning arrangements may be made by voluntary agreement between seed producers and/or plant breeders or included in seed production legislation.

Zoning regulations may call for the registration of any seed or market crop in a specified area regardless of the purpose for which it is produced; this is especially likely for crop species which flower in the course of the development of the marketable vegetable crop. In the USA, sweetcorn seed is produced in Idaho where it is isolated from maize with which it freely cross-pollinates (Delouche, 1980).

The concept of zoning could be applied to the production of GMO seed crops although more care would have to be taken with all compatible plant species in the zone in addition to the seed crops.

Discard strip technique

According to Dark (1971), the pollen concentration in the air over a field of a wind-pollinated crop increases from the windward edge downwind with a tendency to decline again at the lee edge. Therefore, during the period of anthesis when the wind would have been blowing from every direction, a strip around the field's perimeter would have received relatively little of the crop's own pollen and there would have been maximum concentration in the centre. The marginal strip is important in the production of genetically pure seeds. When a cloud of contaminant pollen passes over the field, a small number of pollen grains will drop out at random. Those falling over the centre of the plot will compete with the relatively high concentration of the crop's own pollen and have a lower chance of fertilizing, whereas those which fall on marginal areas will not have so much competition and will therefore have a higher chance of fertilization. Thus, the seeds from a 5 m wide strip around the perimeter of the plot are harvested separately and, according to the genetic quality found when testing a sample by growing it on, they can either be destroyed or placed in a lower category. The bulk of the seed will come from the inner area of the plot and can be kept as a separate seed lot.

Field or plot shape

Most pollen contamination of either wind- or insect-pollinated crops occurs around the perimeter of the plot or field. Therefore, if the area of each crop is kept as near as possible to a square then fewer seeds are likely to have been produced as a result of undesirable pollen and a minimum amount of seed is involved, if the discard strip is applied.

Cultivar Maintenance

During the course of cultivar maintenance or multiplication to increase the quantity of available seed, it is necessary to ensure that the product will be 'true to type'. This trueness to type infers that the plants grown from the new seed generation do not differ significantly from the cultivar's description. The loss, or deterioration, of a cultivar's unique characters is sometimes referred to as 'loss of trueness to type' or 'running out'. Trueness to type may be referred to as the 'cultivar purity standard' and certifying authorities clearly state the maximum distinguishable off-types. Only seed stocks conforming to a cultivar description should be certified.

The seed crop is inspected at stages to ensure that any undesirable plant material is removed as far as possible before pollination and subsequent seeding takes place. This process of removing 'off-types' is frequently referred to as 'roguing'. The roguing of seed crops for maintaining genetic purity has been reviewed by Laverack and Turner (1995).

The generally accepted roguing stages for individual vegetable seed crops are given in Chapters 7–16, although plant breeders who are responsible for the maintenance of individual cultivars or certification authorities responsible for specifying crop inspections for certification purposes will produce their own criteria for different stages. The level of uniformity is usually higher for the predominantly self-pollinated crops than for the cross-pollinated species.

The intensity of roguing or selection during seed multiplication can have a considerable effect on the genetic quality of successive generations. Johnson and Haigh (1966) demonstrated that if selection standards were relaxed in three successive open-pollinated generations of Brussels sprouts (a cross-pollinating crop), there could be an increase in the proportion of off-types. Dowker *et al*. (1971) investigated the effect of selection on bolting resistance and quality of red beet and found that the type of selection imposed during multiplication and the season during which selections were made produced seed stocks with bolting differences in subsequent generations.

There are different methods of ensuring that the most desirable plants in a population are used for the maintenance of seed stocks; these are 'positive selection' and 'negative selection'.

Positive selection

This system may also be referred to as 'mass selection'. The plants are normally selected to a very high standard while still in their vegetative stage. The percentage of plants selected by this method depends on the observed degree of variation in the crop, but the selection of approximately 10% from the overall population is seldom exceeded. It is most commonly used for maintaining breeders', pre-basic and basic seed stocks. Selected plants are usually indicated initially with a cane and may be evaluated more than once depending on when the desirable characters are best observed. In biennial crop species, it is customary to lift the plants and replant them in a block. In some cases, this is done in a structure, cage or isolated area. For root and bulb crops such as carrots and onions, this selection is made at the end of their first season. It is important that the selected plants are moved to their isolated flowering quarters before the remaining unselected plants start flowering. If the selected plants are to remain *in situ* then the unselected plants are removed and discarded before they commence flowering. The seeds from a positive selection are harvested in bulk unless 'progeny testing' is to be done.

Negative selection

This system is frequently referred to as 'roguing'. It is the method most frequently used for open-pollinated seed crops in the final stages or multiplication. The plants that are observed to be within the limits of the cultivar description are retained for seed production, usually *in situ*, while only those plants that are considered to be atypical of the cultivar are removed. Although the percentage of plants 'rogued' out differs according to the species' method of pollination and purity of the seed stock, the proportion of plants removed may be approximately 20%. If 20% are removed by negative selection, this could also be regarded as a positive selection of 80%. Thus, the system of negative selection provides a much lower selection pressure than positive selection and is normally used in the final multiplication stage of an open-pollinated seed crop. The resulting seed crop is bulk-harvested.

The range of off-types or rogues identified in a crop produced for seed production may include:

- Hybrids (occurring from outcrossing in the previous generation).
- Mutations (depending on the crop species' mutation rate.
- Plants resulting from admixture and volunteers.
- Plants which are outside the accepted character range for the cultivar.

Guidelines to ensure efficient roguing

The roguing operation is the basis for ensuring that the seed lot is of the highest possible genetic quality. The following points are therefore emphasized in order to achieve an efficient operation:

- Ensure that weed control and crop thinning are done in a timely way, so that neither crop plants nor weeds obscure the crop.
- Make the roguing inspections as soon as the crop has reached the designated stage, *do not delay*, especially if anthesis is about to start.
- In bright sunny conditions, as far as is possible, have the sun behind you, this is especially important where colour is an important character to be observed.
- Start the work early in the day, if the crop is likely to wilt in high temperatures, otherwise important morphological characters may be overlooked.
- Be positioned so that each plant in the crop is seen individually and that there is no underlap of viewing between members of a roguing gang.
- Remove the entire rogue, off-type plant or diseased plant from the field or plot; *do not* allow rejected plant materials to remain on the headland.
- Remove other cross-compatible crop plants or weeds.

Summary of Maintaining Cultivar Purity (Trueness to Type) During Seed Production

- Use of cages and greenhouses, possibility of including pollinating insects.
- Isolation in time and/or distance.

- Zoning, especially cross-compatible wind-pollinated crops.
- Suitable period of rotation between related crop species and suitable time lapse between crops liable to leave undesirable vegetative material or seed in the soil which will produce undesirable volunteer plants within the seed crop – zoning and scientifically agreed isolation of GMOs.
- Follow guidelines to ensure efficient roguing.

Methods and Techniques to Assess Postharvest Seed Lot Quality

Progeny testing

This technique is widely used for testing the material obtained for breeders' stocks but it can be used at any stage of multiplication. It is especially useful for crops which are cross-pollinated.

Selected plants are open-pollinated but the resulting seed from each selection is harvested separately. A small sample from each seed lot is grown on as separate progeny. The performance of each progeny is assessed and the main seed lots from which the satisfactory progeny lines were grown are bulked together and used for further seed production. The individual seed lots from which the undesirable progenies were grown are discarded.

Trial grounds and growing-on tests

The types of trials discussed in Chapter 1 included the cultivar trials, testing for distinctness, uniformity and stability (DUS) and trials by plant breeders to evaluate material in the course of development. In addition to these trials, seed companies will normally grow on a sample of each new seed stock as a check on its genetic purity. This growing-on test (sometimes referred to as 'growing out') is in addition to the standard germination test and any other trials or tests for which samples may have to be submitted. Many companies also include commercial samples from their competitors, especially of new cultivars. In this way, a seed house not only monitors the growing-on quality of its own seed stocks, but is also able to see the cultivars marketed by other companies and new releases from research institutes and elsewhere.

Plant Production

The agronomic methods and systems used to obtain a satisfactory plant stand for seed production are generally the same as for the production of fresh vegetables, and either direct drilling or transplanting is used according to the crop species and local conditions. The methods for individual seed crops are described in Chapters 7–16.

There are two terms that are commonly used by vegetable seed producers according to the method used when referring to biennial root and bulb crops. 'Seed to seed' refers to a crop which is sown directly in its final quarters and the plants remain *in situ* at the end of their first season. With this system in which the plants overwinter without being transplanted, it is not possible for selection or roguing of root characters and it is therefore normally only used for the final multiplication stage of crops such as carrots and onions; the 'seed to seed' method should never be used for the production of the basic seed category, as it does not allow for adequate scrutiny of the morphological characters of individual plants in the mother plant population retained for seed.

The term 'root to seed' is generally used for biennial crops when the plants are to be lifted at the end of their first season and replanted after discarding the undesirable material. In some areas, the overwintering plant material is stored between lifting and replanting.

Some root crops, for example, red beet, are sown later in their first season and transplanted as relatively small plants either before or after the winter. Plants produced for this method are frequently referred to as 'stecklings', and while they are of insufficient maturity when transplanted for all their characters to be assessed, some degree of verification of morphological characters can be made depending on the species.

Mother Plant Environment

The modification of the microclimate within seed production areas was discussed earlier in this chapter. It is also important to ensure that the soil or substrate is not a limiting factor for satisfactory mother plant growth and development. Generally, the same principles of crop nutrition apply for seed production as for the market crop. However, in view of the increased time required to produce a seed crop, the loss of nutrients due to leaching must be taken into account by the application of the appropriate top dressings in the later stages of plant development. There is increasing evidence that the nutrient regime of the mother plant not only influences seed yield but also seed quality. For example, Browning and George (1981a,b) showed the influence of NPK regimes on pea seed yield and quality.

The optimum soil pH and ratios of NPK fertilizers for specific crops, together with important micronutrient requirements, are given for individual crops in Chapters 7–16.

The Use of Pesticides for the Control of Pathogens, Pests and Weeds

Generally, the same control methods are used in seed production as in the production of market crops of vegetables. The range of available pesticides differs from one country to another, but only approved or proven products should be used in seed production, as possible adverse effects of pesticides include the inadvertent killing of pollinating insects, a modification of the seed's potential germination and a reduction in overall seed quality. Little work has been done on this last aspect of the effect of pesticides on seed quality, but Olympio

(1980) showed that in lettuce and tomato some herbicides did not affect seed yield or quality although trifluralin, at the recommended rate, increased tomato seed yield, whereas diphenamid decreased seed quality. Propyzamide applied to lettuce at or after anthesis reduced the seed yield.

Organically Produced Vegetable Seed

The production of vegetable seed for organic farming systems has its own special requirements. Strict adherence to the prevailing specifications for organically produced seed must be adhered to in any country of production for seed to be used for authenticated organic vegetable crops.

Background

The demand for organically produced vegetable crops for the fresh market became popular with consumers towards the end of the 20th century and has continued to increase in the 21st century. The popularity of organic produce increased especially in some North American, European and southern hemisphere markets, although not confined to these areas. The call for organic crops has also included various forms of processed vegetables. The International Federation of Organic Movements (IFOAM), with its head office in Bonn, Germany, has a global role with links to member organizations in some 108 countries. IFOAM liaises with the main international seed organizations, including the Seed Service of FAO and ISF.

Definition and possible future impact

The term 'organic vegetables' refers to the method of production. The fundamental principle is that the crops are produced without the use of specified crop protection chemicals or inorganic fertilizers on land that is defined as suitable; (i.e. approved by an overseeing organization).

The development of market outlets for organic produce has created a demand for organically produced vegetable seed, which is produced under similar conditions as the organic ware crops. Some governments have aimed to increase the percentage of crops produced organically, which in turn is expected to increase the market needs for organic seed.

However, vegetable producers in developing countries may face difficulties in marketing their produce in developed countries if their seed specifications do not meet the prevailing criteria.

Legal framework

As with all official schemes, and with consumer protection in mind, there needs to be an overseeing organization or agency in individual countries to monitor the authenticity of crops. In the European Union (EEC), the main starting point is the EC Directive 2092/91, which stipulates that the seed used for organic crop production should also be produced in an organic regime.

In the UK, the Soil Association has taken the regulatory role and responsibility for authenticating the seed produced. In other European countries, similar organizations have taken on the monitoring tasks.

In the USA, all producers and handlers of organic crops are required to be approved by a USDA certification agent to label or sell a crop as 'organically grown'; the regulations also include standards for seed sources for organic crop production (Bonina and Cantliffe, 2005).

In the early stages of its implementation, there were derogations in place which allowed producers of vegetables destined as organic to use non-organically produced seed ('orthodox seed'), if the seed of a required cultivar was either not available or unavailable in sufficient quantities; this created planning difficulties for the seed producer.

Organic seed and seed suppliers

Initially seed producers and suppliers need to know if there is a market for their products. There are several factors to be considered in the case of organic seed. These can include the following:

- The cultivars to include in seed production programmes, as this will very much depend on the cultivars opted for by the organic vegetable producers.
- Generally, the vegetable producers will choose cultivars according to their market requirements, e.g. suitability for fresh produce, pre-packaging or processing.
- The growers' organic production regimes, including which cultivars have better pest and pathogen tolerance or control when grown in organic regimes.
- The estimated quantities of seed required for the required cultivars.
- The commercial risks in opting for production of organic seed.
- The feasibility of producing organic seed with only allowable materials, especially for the biennial crop species which have a longer exposure period to pests and pathogens in the field.
- Finding alternative viable methods of weed, pest and disease control during seed crop production.
- The seed producer has to decide on a plan when embarking on the production of organic seed that will satisfy the requirements stipulated by the monitoring authority.
- A suitable plan for control of pests, pathogens and weeds will have to be made, and it will very likely imply the use of biological controls possibly along with integrated control systems.
- Additional considerations by the seed producer may include: likelihood of a derogation of a cultivar by the controlling authority and therefore possible loss of sales potential, if the seed producer has speculated on such a cultivar.
- Many seed producers have already successfully incorporated organic vegetable seed production programmes into their businesses; while others will need to develop acceptable protocols, if they plan to.

- It is generally accepted that the production costs of organic seed are higher than for orthodox seed, due to lower yields in some species, depending on the crop.
- Many of the pre-sowing treatments to control, or partly control, pre- and post-emergence of pests and pathogens in orthodox seed production are not allowable for organic seed. The later section in this chapter includes pre-sowing treatments, some of which are long standing, and also newer techniques allowed for organic seed, some of those listed which include chemical treatment would not normally be allowable.
- It is essential that the various monitoring authorities allow seed producers the same derogations as the organic vegetable crop producers have.

Research and discussions related directly to the needs of organic seed production

It has been generally demonstrated and agreed that there is a need for further research and development into the requirements of organic seed production.

The Conference Report (IFOAM, ISF and FAO, 2004) stated that two legitimate realities exist, that is:

1. Large-scale vegetable growers, aiming to supply supermarkets in their home countries and export markets where there is heavy reliance on modern cultivars, including F1 hybrids.

2. Farmers groups in developing and developed countries, where local markets are supplied often using 'local varieties', from community-based seed production systems and in some instances involved in participatory plant breeding.

Both of these groups have some different requirements regarding their objectives.

An example of an earlier investigation into feasibility of producing organic seed in England

Work at Horticultural Research International in England (DEFRA, 2003) examined the organic seed production of three biennials, i.e. parsnip, winter cauliflower and bulb onions, also a small range of annual vegetable species. The experiment demonstrated that while it is possible to produce onion seed in tunnels under UK conditions, the incidence of *Botrytis allii*, (onion neck rot), remained a problem. This work also showed that while seed quality differences were small, the management costs can be high. From this example and discussions held at the International Conference referred to above, suggestions for some further research and development can be identified. These include:

- Identify new seed production areas which can be shown to have a reduced incidence of those pests and/or pathogens which are difficult to control in traditional seed production areas, especially biennial seed crops.

- Evaluate existing cultivars to determine their suitability for crop production under organic production protocols.
- Identify plant breeding opportunities for the development of cultivars that fulfil the needs of organic vegetable production protocols.
- Participatory plant breeding can offer an opportunity for close liaison between breeders and organic growers for the further development of cultivars most suited for organic production. For example Lammerts *et al.* (2005) have discussed the need to broaden the genetic base of crops such as onion when breeding to meet the cultivar requirements of organic growers.
- The biological control systems can be used to great advantage in some situations during seed production.

Nutrition and guidance on organically produced seed

Crop rotation and approved organic and mineral materials are a part of the concept of organic production. However, from the point of view of seed production, care must be taken to ensure that undesirable crop or weed seeds and propagules are not introduced with uncomposted organic materials. Generally, synthetic crop protection materials should not be used. The control of pathogens, pests and weeds has to be achieved by appropriate husbandry (including choice of cultivar, rotation, methods of encouraging natural predators and timely cultivations). With the advent of organic farming systems and increased demands for authentically produced organic vegetables (in addition to other organic products), there are various monitoring and advisory agencies. For example, in the UK, further information may be obtained from the Soil Association Producer Services and the UK Register of Organic Food Standards (UKROFS). The National Institute for Agricultural Botany (NIAB) has included vegetable cultivar trials on UKROFS-approved organic farms and published lists of vegetable cultivars suitable for organic production in the UK (NIAB, 2007).

Summary of Organic Seed Production

- Organically produced seed concept, legal framework.
- Feasibility, role of producers of organic seed.
- Production, requirements, restrictions and seed suppliers.
- Nutrition of mother plants with organic fertilizers and manures.
- Care required with organic manures and fertilizers to produce a satisfactory seed crop.
- Different requirements of large-scale commercial producers compared with farmers groups in developing countries.
- Organic seed is subject to the same seed quality regulations as orthodox seed, in addition to the prerequisites stipulated for its authentic production.

Pre-sowing Seed Treatments

There are several pre-sowing treatments that are used for vegetable seeds, including the application of pesticides for the control of seed- or soil-borne pathogens, modification of seed shape or size and pre-germination before sowing.

Seed treatments for the control of pathogens

The range of vegetable seed treatments and their methods of application for the control of fungi, bacteria and insect pests have been reviewed by Maude (1986). The available treatments range from the application of chemicals to the seed as dusts or slurries to the application of heat via hot water, dry heat or steam–air mixtures. Developments have included the use of antibiotic solutions and more recently the addition of chemicals to hot water.

The need for organic seed treatments to replace the use of disallowed chemical treatments, which have become the norm with orthodox seeds, has been indicated by Groot *et al.* (2004) who cite the possible use of essential oils and organic acids. The evaluation of hot, humid air seed treatments and fluidized beds have been described by Forsberg *et al.* (2000).

The methods for the application of pesticides to a wide range of seeds have been reviewed by Jeffs (1986). Seed-borne nematodes (eelworms) in onion and leek seeds have been controlled in the past by fumigation with methyl bromide (Baker, 1972). The use of methyl bromide as a soil or seed fumigant has now been phased out in many countries.

The main seed-borne pests and pathogens

The main seed-borne pathogens with the common names of the diseases they cause are given in Chapters 7–16 for the seed crops discussed. It must be emphasized that many of the pathogens listed can be transmitted by other vectors and means. The nomenclature of the pathogens is in accordance with *An Annotated List of Seed-borne Diseases*, published by the International Seed Testing Association (Richardson, 1990); the list includes fungi, bacteria, viruses and nematodes which affect seed or seed assessment.

Modification of Seed Shape and Size: Seed Pelleting and Protective Treatments

Pelleting

Pelleting facilitates the manual and mechanical handling of seeds which are either small or awkwardly shaped. Precision drilling direct in the field or sowing in soil blocks using pelleted seed prior to transplanting can ensure a high degree of accuracy regarding seed placement. Individual seeds are encased in an inert material such as montmorillonite clay. The implications of seed pelleting with special reference to seed quality were reviewed by Tonkin (1979). The pelleting material can also be used to incorporate pesticides or form a coating to seed dressings. The technology for pelleting vegetable seeds was gained largely from developments with pelleting sugarbeet seed (Longden, 1975). The pelleting of seed is usually done by specialist companies using proprietary processes at the request of individual seed companies. The development of commercial seed treatments has been discussed by Halmer (1994).

Coating

Seed coating, or film coating, is a technique by which additives such as pesticides, nutrients or nitrifying bacteria can be applied to the external surface (i.e. the testa) of the seed. But in contrast to pelleting, the coating conforms to the seed's shape and does not normally significantly modify the seed's size. An air-suspension technique originally used in the pharmaceutical industry is used (Wurster, 1959). A film forming polymer containing the required active ingredient is sprayed on to the seeds while they are suspended in a column of air which is either heated or unheated. A colouring agent is usually incorporated at the same time and the coating materials dry very quickly, resulting in free-flowing coated seeds. The coating of horticultural seed with films has been described by Robani (1994); Scott (1989) has discussed the effects of seed coatings and treatments on plant establishment.

Organic Seed Production and Pre-sowing Treatments Which Include Pesticides

Many of the very successful seed treatments developed over the years prior to the start of organic production have to be evaluated. For example, the thiram treatment developed by Maude and Keyworth (1967) was widely adopted but cannot be used in the processing of seed intended for the organic market. Traditional treatments such as hot water treatment (HWT) have come back into wider use as also have other allowable methods; in addition, there are some interesting new developments which could also be effective for the orthodox crop producers. The wide range of treatments for vegetable seed was reviewed by Maude (1986) and includes several methods suitable as pre-sowing treatments for organic seed. The challenges for the seed coating and pelleting industries have been reviewed by Legro (2004).

Postharvest and pre-sowing treatments for organic seed

There is a need for the organic seed producers and suppliers to consider the availability and application of appropriate postharvest treatments requested by purchasers which are allowed by the controlling authority where the seed will be marketed or used.

The minimum requirements for seed quality, as required by seed legislation remain the same for all seeds regardless of production protocol. However, organic growers may well require more emphasis on monitoring seed for presence of GMO material, from both admixture and genetic contamination.

The biological control systems can be used to a great advantage in some situations during seed production.

Hydration Treatments

There are two main types of hydration treatment for seed, i.e. priming and moisturization. Taylor (1997) has reviewed the various treatments in relation to the storage, germination and quality of vegetable seed.

Seed priming

The priming of seed is a pre-sowing treatment which aims to have all seeds in a given lot on the 'brink' of germination prior to sowing. Thus, while an initial seed lot may contain a population of seeds with differing potential durations to germination, the technique of priming will ensure that each seed in the lot has reached the same stage of germination prior to sowing. Priming, which is in two basic stages, was described by Heydecker (1978) and Heydecker and Gibbins (1978). The technique has been further developed and several materials have been used for the preparation of the priming solution (Khan, 1992). Polyethylene glycol (PEG) has been the main material used to control seed hydration levels in the priming process. The process of seed priming has been used for a wide range of vegetable crop species, for example, Dearman et al. (1986) have studied the effects of osmotic priming and ageing on onion seed germination. Primed seeds may be dried back or sown by 'fluid drilling' techniques.

Rowse (1996a) has described the concept of 'drum' priming. The commercially adopted system (Rowse, 1996b) involves the hydration of seeds to a predetermined water content over a 1–2-day period by placing them inside a rotating drum into which water vapour is released. The drum is mounted on an electronic balance linked to a computer which monitors the seeds' water content and controls water vapour production, so that the seeds do not become visibly wet. The seeds are then incubated in a rotating drum for a period of up to 2 weeks before being dried to the moisture content suitable for their storage. The drum priming method avoids the problem of PEG disposal which occurred with priming commercial seed quantities.

Parera and Cantliffe (1994) have reviewed pre-sowing priming of seeds. Large-scale seed priming techniques and their integration with crop protection have been discussed by Gray (1994). The effects of priming to germination of unprimed seeds of carrot, leek and onion have been evaluated by Drew et al. (1997) who suggested that slow-germinating seed lots benefit more from priming than faster ones, and that the effects of priming on germination rate may be predicted from a standard test of unprimed seeds, but that storage potential of primed seed is limited.

Pelleting

It has been demonstrated that seeds with low moisture contents may be prone to both mechanical damage (Bay *et al.*, 1995) and low germination rate (Taylor *et al.*, 1992). Wilson and Trawatha (1991) have described and discussed moisturization as a method of enhancing the emergence of bean seedlings. Taylor (1997) indicates that in this process the seed moisture content is adjusted to approximately 12% and suggests that while the moisture content can be adjusted prior to packaging, too high a level is likely to lead to an acceleration of seed ageing.

Combining hydration and coating

Taylor *et al.* (1991) have described a system for upgrading *Brassica* sp. which combines hydration and coating of seed. The seed is first hydrated in water or primed; during this process, sinapine leakage commences from the non-viable seeds; the seeds are then coated with an adsorbent which retains the leakage on individual seeds. Following drying, the seeds are exposed to UV light and sorted according to their fluorescent coatings. The non-fluorescent fraction produced an increased proportion of normal seedlings than the control, while the fluorescent fraction contained a higher proportion of dead seeds, or seeds which subsequently produced abnormal seedlings. Taylor *et al.* (1993) have described a sorting method for *Brassica* sp. using a colour sorter with a UV light.

Fluid Drilling

The term 'fluid drilling' is used when seeds that have been pre-germinated are sown. A range of machines has been developed for dealing with different scales of operation. The topic of fluid drilling is considered to be beyond the scope of this book. The techniques and prospects for fluid drilling of vegetable crops were reviewed by Salter (1978a,b).

Summary of Postharvest and Pre-sowing Seed Treatments

- Pelleting and coating, advantages, materials, methods and crop protection materials included.
- Organic seed and pre-sowing treatments, applications and restrictions also for postharvest treatments.
- Hydration treatments: unifying stage of germination, materials and methods.

Further Reading

Organic seed production

IFOAM, ISF and FAO (2004) *Proceedings of First World Conference on Organic Seed*. FAO, Rome.

Maintaining cultivar purity

Bateman, A.J. (1946) Genetical aspects of seed-growing. *Nature* 157, 752–755.
Faulkner, G.J. (1983) *Maintenance, Testing and Seed Production of Vegetable Stocks*. Vegetable Research Trust, NVRS, Wellesbourne, Warwick.
Haskell, G. (1943) Spatial isolation of seed crops. *Nature* 153, 591–592.

4 Harvesting and Processing

The period of anthesis of many of the cultivated vegetable crop species is relatively long, with successive flowers on complex inflorescences, resulting in a long period of fruit ripening and seed maturation. In many, e.g. *Brassica* species, lettuce and okra, there is a strong tendency for the earlier maturing seeds to ripen and drop before later ones have developed. This loss before harvesting is often referred to as 'shattering' or 'shedding'. There has been an active interest in the use of polyvinyl acetate sprayed on to seed crops in early maturity to prevent shattering. The material acts as glue which dries on the plant and prevents shedding of the earlier matured seed while the later seeds continue to develop.

Other sources of seed loss before harvesting include birds, small rodents and inclement weather. There is a wide range of bird-scaring devises and methods used throughout the world to prevent loss including audio methods (e.g. methane 'bangers' or boys with rattles) and visual methods (e.g. scare crows and aerial balloons). Netting is successful for the protection of small-scale seed production areas but is impractical over large areas.

Lodging is a term used to describe the collapse of a crop before cutting or harvesting. Crops for seed production are especially vulnerable from the start of anthesis when the extra weight of the inflorescences or seed heads makes the mother plants top-heavy.

In addition to the susceptibility of specific crops such as lettuces to lodge after seed head emergence, several cultural and environmental factors can contribute to the incidence of lodging: these include wind, high nitrogen regimes during early plant development, heavy rain which can either weigh plants down or reduce efficiency of root anchorage and straying animals.

Once lodging has occurred, the plants tend to deteriorate and, depending on their development stage, may not regain their normal vertical posture. This can result in a poor micro-environment in the crop canopy, insufficient pollination; and in relatively wet seasons, or areas, the seed quality may deteriorate with reductions in seed yield, viability and vigour.

Stage of Harvesting

The effects of harvesting stages of peas and carrots have been reported by Biddle (1981) and Gray (1983), respectively. Biddle's work with peas emphasized the need for pea seed moisture content to be between 30–44% for threshing in a pea viner and at approximately 26% when combine harvested. Gray's work with carrots clearly showed a relationship between embryo size not only with position on the plant but with seed maturity. The effects of carrot umbel order and harvest date on seed variability and seedling performance were reported by Gray and Steckel (1983). While the above examples of effects of harvesting stage on seed quality demonstrate the complexity of the topic, there is little published work on these aspects in relation to other vegetable seed crop species which are not in *Apiaceae* and there is still scope for further research.

In addition to the seed's development and ripening rate characteristic for the species, ripening is accelerated by relatively high temperatures, a low soil moisture level and a low relative humidity. Conversely, the rate is reduced by the reciprocal of these factors. The environmental effects before harvesting on seed viability were reviewed by Austin (1972). The ripening process is interrupted if the seeds are harvested too early and seed quality may be adversely affected.

Generally, the later the harvest, the higher the seed yield, but as shown in Fig. 4.1, losses increase as harvesting is delayed once the optimum percentage of seeds reaches maturity. Thus, for any individual crop, the ideal harvest time

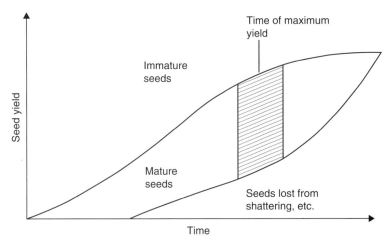

Fig. 4.1. Schematic representation of the interaction of seed maturity, potential yield and loss by shattering and birds.

is immediately before the loss of mature seeds exceeds the amount of seeds yet to reach maturity.

The incidence of shedding from ripe material is increased during dry weather. Crops which are particularly prone to shedding during cutting should be handled at times of comparatively high relative humidity. In arid areas, this can be early in the day when the effects of overnight dew are still effective, after rain or even following irrigation.

Although seeds that are produced in wet or fleshy fruits, such as cucurbits, peppers and tomatoes, are not normally subject to all the effects of environment as are dry-seeded vegetable species, it is generally accepted that the seeds should be allowed to develop fully in the fruit before extraction. Ideally the fruit should remain attached to the mother plant, but work by Cochran (1943) showed that the stage of fruit maturity at the time of harvest had a significant effect on potential germination of *Capsicum frutescens* (pimento pepper): seed extracted from 30-day-old fruits gave approximately 5% seedling emergence when tested in compost, but fruit picked at the same age and stage which were kept for a further 30 days before seed extraction produced seeds which gave a seedling emergence of 95%.

Summary of Pre-harvest Causes of Seed Crop Quality Reduction and Yield

Loss of seeds before and during harvesting

- Shattering, birds, lodging, vulnerability of seed-laden mother plant from weather conditions or other hazards.
- Lodging, possible effects on seed yield and quality.
- Effect of exact harvest stage.
- Balance between immature seeds yet to ripen against subsequent loss, and reduction in yield and quality to achieve maximum yield of high quality seed.
- Effect of weather on yield and quality.

Types of Material to be Harvested

The types of vegetable seed material to be harvested can be broadly classified into three groups, i.e.:

- Dry seeds (e.g. brassicas, legumes and onion).
- Fleshy fruits which are usually dried before seed extraction (e.g. chillies and okra.
- Wet fleshy fruits (e.g. cucumbers, melons and tomatoes).

In some crops, the method of harvesting and extraction depends on the scale of operation, the stage of multiplication (i.e. the seed category being produced) and on any links with food processing. For example, pepper seeds are extracted from the wet fruits, if large quantities are produced in conjunction with a dehydration plant, but if only small quantities are produced the seeds are extracted from relatively dry fruits after fully ripening.

Harvesting Dry Seeds

The methods of harvesting dry seeds include:

- Cutting off individual seed heads.
- Cutting the majority of the plant and leaving the material in windrows to dry further before seed extraction.
- Using a combine harvester which extracts the seed from the seed heads and separates it from the remainder of the plant debris (i.e. the 'straw') in one operation.

In the first of these two methods (i.e. removal of individual seed heads and cutting for further drying), the seed is subsequently separated from the 'straw' by threshing.

Harvesting by hand

Hand harvesting is still done for very high-value seeds, when the total area to be harvested is very small (e.g. breeders' and basic seed) or in areas where there is adequate hand labour available. Seed heads, dried fruits or other forms of modified inflorescence containing the mature seeds (e.g. sweetcorn cobs and leek flower heads) are picked or cut with knives or secateurs into baskets or other suitable containers. In some crops that are cut by hand, a larger part of the plant is removed with the seed heads. This is achieved, for example, with radish, lettuce and brassicas, by the use of knives, machetes, sickles (reaping hooks) or in the case of some crops, such as peas, by pulling up the whole plant.

Hand-harvested material is usually either dried further on tarpaulins or other suitable sheets, or placed in suitable buildings or structures on clean concrete floors with a smooth surface, or in airy racks or boxes.

Desiccants

In some seed crops, it is necessary to dry-up the plant material when the seed is approaching maturity. This facilitates mechanical harvesting of the seed by reducing the amount of plant debris, increases the rate of seed drying, brings all the seed material within a closer range of moisture content, avoids loss from windrows and will also to some extent control weeds. Materials that are used for this purpose are known as desiccants of which an example is diquat. Although desiccants have been used on crops such as carrot, chicory and dwarf beans, the main vegetable seed crops to which they are applied are *Beta* species. The use of dessicants is relatively expensive; there is a possibility that they may cause seed dicoloration and a reduction in potential germination.

Mechanized cutting

When the cutting operation is mechanized, the material is either left *in situ* by a machine with a cutter bar or a machine which places the material in windrows. Machines capable of this operation have a draper belt in addition to the cutter bar, and the cut material is carried under the machine and deposited in a swath. The swaths can either be turned into windrows or left *in situ* to dry according to the density of the material and the rate at which they will dry in the field.

Combine harvesting

This joint operation is done by combine harvesters which cut and thresh the material in one process; however, it is important that the standing material is uniformly ripe. Combine harvesters can be used to harvest seed from a large-scale seed crop, such as lettuce which has lodged. These machines can also be used as stationary threshers or for collecting and threshing material already cut and which has been left in windrows for further drying, in which case a pick-up head or reel is fitted.

Threshing

Dry seeds are removed from the mother plant material by flailing, beating or rolling. It is important to ensure that unnecessary fragmentation of plant material does not produce debris which is either difficult or costly to separate from the seed sample during subsequent processing; it is also important to avoid mechanical damage to the seeds.

Hand threshing

Hand threshing is a relatively cheap method for small seed lots and is still used in some countries for large seed lots where labour is cheap. Several hand methods are possible including rubbing, beating the material against a wall or the ground, or flailing.

Rubbing seed materials with a presser in an open-ended trough lined with corrugated rubber is very suitable for material in pod-like structures such as brassicas and radish.

Seeds which have been hand-threshed are usually still mixed with the plant debris, and further separation is done by winnowing and/or sieving.

Machine threshing

The main feature of threshing machines is a revolving cylinder in a concave. The cylinder is driven by a motor or engine and is capable of reaching 1200–1500 revolutions per minute (rpm). While the potential of these high rpms is

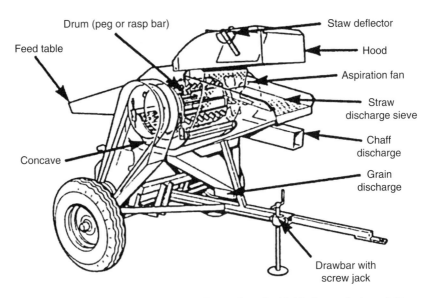

Fig. 4.2. Diagram of a mobile thresher. Reproduced with kind permission of Alvan Blanch Ltd, Chelworth, Malmsbury, UK.

suitable for agricultural grain, speeds of 1100 rpm are used for small-seeded vegetable species, and as low as 700 rpm for the large-seeded legumes. There are stationary and mobile threshers manufactured.

Both the cylinder and the concave have spiked steel teeth, angle bars, rasps or rubber bars; threshers with bars are more frequently used for vegetables than models with teeth. The concave opening is adjusted to allow the free passage of seeds which are collected in a container, usually below the drum. Some threshers have interchangeable concaves for use with large-seeded legumes.

Some types of thresher incorporate sieving, screening or aspirating components to assist in the initial separation of seeds from plant debris. These modifications or additions can include the possibility of adjusting cylinder speed, cylinder clearance, concave mesh, air flow and screens. These refinements are essential for dealing with a range of vegetable species.

The possibility of surface and sub-surface damage to seeds is increased, if the cylinder speed is too fast, the cylinder clearance is too narrow or the mesh of the concaves is too small. Figure 4.2 illustrates a mobile thresher.

Seed Extraction from Wet or Fleshy Fruits

Extraction of seeds from wet or fleshy fruits is either done by hand or by specially designed machines. The specific extraction methods, machines and any special post-extraction seed-cleaning requirements are described for vegetable seed crops in *Cucurbitaceae* in Chapter 10, and *Solanaceae* in

Chapter 12. Generally, the wet extraction machines work on the principle of fruit maceration followed by separation of crop seed from the fruit pulp with other fruit debris. The performance of an axial-flow vegetable seed extractor suitable for tomato, aubergine, watermelon, muskmelon and pumpkin has been reported by Kachru and Sheriff (1992).

Summary

Harvesting and seed extraction

- Types of fruiting body from which to extract seeds, i.e. dry seed heads, fleshy fruit and wet fruit (*see* also Chapters 10 and 12 on *Cucurbitaceae* and *Solanaceae*).
- Threshing and seed extraction systems.

Combining harvest and seed extraction

- Types of fruiting body from which to extract seeds, i.e. dry seed heads, fleshy fruit and wet fruit.
- Area to harvest and scale of operation: choice of hand or machine.
- Threshing, small scale or amounts by hand, large areas by machine.
- Combine, reap and thresh simultaneously.
- Use of combine harvester as stationary thresher.
- Seed extraction of wet or fleshy fruit described in Chapter 10 (*Cucurbitaceae*) and Chapter 12 (*Solonaceae*).
- Care with threshing and other processing to avoid damage to seeds.

Seed Processing

The term 'seed processing' is used by the seed industry to include a wide range of operations to improve or 'upgrade' seed lots after threshing or extraction. The objectives of processing may include removal of a wide range of materials including:

- Plant debris.
- Non-plant material (e.g. soil or stones).
- Seeds of other crops and weeds.
- Seed appendages which would otherwise interfere with the free running of the seeds.
- Damaged and discoloured seeds.
- Seeds which are outside the accepted size or density tolerances of the seed lot.

Winnowing

After threshing, the dried seeds can be separated from less dense or lighter debris by air movement, usually referred to as winnowing; the operation can be performed by hand or by machine.

Pre-cleaning ('scalping')

During pre-cleaning, the bulk of plant debris and any other non-seed materials are separated by vibrating or rotating sieves. In some species, seed clusters are also separated during pre-cleaning. The pre-cleaning machines usually have an air flow to remove materials lighter than the seed. Seed lots are usually pre-cleaned before drying. Figure 4.3 illustrates a pre-cleaner.

Basic cleaning

The main cleaning operation is generally referred to as basic cleaning. During this stage, all materials should be removed from the seed crop with the possible exception of contaminants or materials of any origin that are intended to be absent from the final seed lot, but which will require a further specific separation process.

The simplest form of basic cleaning is the use of screens (sieves) which separate according to seed size. Seed-cleaning machines are designed to combine air and screens; they are known as 'screen and air cleaners'. These machines incorporate motorized fans which force air through the cleaning chamber or draw air out. They usually have at least two vibrating screens. The principle of their functions is that the air (aspiration) system separates by seed and particle weight, while the screens separate by size (width and thickness).

Seed cleaning with a screen and air machine can normally achieve a very high level of purity in the seed lot. It is the main method of seed cleaning and should be done as skilfully and efficiently as possible.

Fig. 4.3. A pre-cleaner. Reproduced with kind permission of Alvan Blanch Ltd, Chelworth, Malmsbury, UK.

Separation and Upgrading

These are normally the final processes which improve the mechanical purity of the seed, and may be done to remove a specific contaminant from the seed which was not achieved by the screen and air-cleaning operation. In the case of removal of appendages attached to the crop seed the operation is done earlier in the processing sequence.

ISTA (1987) provided a very comprehensive account of the cleaning of horticultural seeds with detailed descriptions of a wide range of machines.

The more important seed-processing machines used for vegetable seed processing in addition to screen and air cleaners are described below.

Maize sheller

Sweetcorn seed is usually removed from the cobs by subjecting them to a shelling action. As the cobs pass through a drum, which has a rotating beater with pegs, the seed is separated through a concave screen. Some machines incorporate fans to remove dust and debris from the cobs. The design of individual machines depends on the potential throughput, and whether or not it is for fixed installations or mobility, see Fig. 15.1. Care must be taken to ensure that the type of machine used does not damage the seed.

Debearder

Some seed lots require further attention immediately following threshing or extraction before they are in a suitable physical state for processing in a screen and air cleaner. Carrot and dill seeds have appendages that result in the seeds adhering to one another to form clusters, which interfere with the free flow of the final seed lot. The seeds of some crops, such as tomatoes and cucumbers, which have been extracted from wet fruit may contain clusters or 'doubles' of dried seed which has to be separated prior to further processing. The removal of seed appendages and separation of clusters are usually achieved by passing the seed lot through a debearder, this machine may also be referred to as a 'de-awner'.

The debearder, which is so called because they are used extensively for debearding barley seed, is essentially a drum with horizontal arms rotating inside. The arms rub the seeds against the drum and thereby remove appendages or separate clusters.

Spiral separator

The spiral separator, which has a minimum of two spirals around the vertical axis, is used for separating non-spherical or irregular-shaped seeds from a round-seeded species, for example, to separate *Brassica* seeds from other material and broken seeds.

The seed material to be processed is put at the top of the inner spiral column; the more rounded seeds have a greater velocity and travel down the outer spiral while the flatter and irregular-shaped seeds travel down the inner spiral. The separated lots are collected at separate outlets or 'spouts' at the bottom.

Disc and cylinder separators

Generally in the modern seed-processing industry, these types of separators have replaced the spiral separators. They operate on the principle that one fraction of a seed lot is picked up in small depressions in a disc or inner surface of a cylinder, while the other fraction remains loose and is thereby separated. An example of their application is to separate pieces of stem from lettuce seeds and for grading some seed crops.

Gravity separator or gravity table

These machines separate according to specific gravity, and are capable of separating good seeds from other materials, including seeds which are mechanically damaged, diseased, light-weight, sterile or insect-damaged. The seeds are fed on to a vibrating and sloping deck above a plenum chamber; air passes through the deck which is covered with either a porous cloth or fine wire mesh. The deck's vibration, and lateral and lengthways slopes are all adjustable, as also is the air distribution. There are two basic types of gravity separators, according to whether the deck is rectangular or triangular. The different grades of material move in different directions on the deck and are collected separately. The separation of the fractions into horizontal layers during the machine's operation is usually referred to as stratification.

Magnetic separator

This machine is used for separation and depends on the differing surface characters between the two fractions. The seed lot to be upgraded is first treated with iron dust, moistened with water or oil, which adheres to the rough-coated fraction (e.g. seeds of the common weed *Galium aparine* contaminating a sample of radish or *Brassica* seeds). The material is then passed over magnetized rollers and the clean seed is collected at the side while the magnetized material which has picked up iron dust is brushed off separately.

Electronic colour separator

These machines are used for the separation of 'off-colour' seeds from an otherwise clean seed lot. Examples of their use include removal of individual discoloured or stained pea or bean seed, or any seed which is infected with halo blight. The seeds are fed by belt, gravity or roller past a photoelectric cell which triggers a jet of air to remove the off-colour seed from the main seed lot. Both monochromatic and bichromatic instruments are available to enable different colours as well as different intensities of the same colour to be detected and removed.

Precision air classifier

This type of machine separates materials with a range of sizes and specific gravities by floating them through a rising air-stream. The machine works on the principle of differential specific gravities between fractions which can be separated by fine adjustments to the machine.

Brushing machine

Brushing machines are used when a special cleaning or separation is required such as detaching carrot seed from dry umbels or the removal of stalks and dried inflorescences from lettuce seed. Brushing machines are also useful for cleaning small seed lots, for example, the relatively small amounts of seed harvested from selected plants for the production of basic seed.

The machine consists of a set of brushes which rotate against specially prepared cylindrical mesh surfaces. The separated seeds are brushed through the mesh and the remaining debris is collected separately and discarded.

Needle drum separator

Seed of peas and beans which have damage caused by weevils can be separated from sound seeds by a needle drum separator. Separation is made by a series of needles housed on the inside of a rotary drum. Seeds are transferred to the revolving drum, those seeds which have weevil holes become attached to the needles and are separated from the sound seeds.

Summary: Seed Processing

The range of operations may be subdivided into four basic groups, i.e.:

* Winnowing.
* Pre-cleaning (often referred to as 'conditioning' or 'scalping').
* Basic cleaning.
* Separation ('upgrading').

The unit for seed processing consists of a range of machines some of which are suitable for a wide range of crops while others are designed for specific crop genera, e.g. the maize sheller, or specific problems arising in quality control, e.g. the needle drum separator.

The separation of seeds from other materials is based on physical differences such as size, shape, length, density, surface texture, colour, affinity to liquids or relative conductivity.

Choice of machine depends on:

- The crop species.
- The stage of processing.
- Problems with an individual seed lot.

Management and Organization of Seed Processing

It is extremely important that machines are cleaned between seed lots. This is usually done by vacuum cleaning the removable parts such as screens and also the interiors of machines. Floors and other areas around the processing machines must be kept clean and free from debris.

The processing machines should be maintained according to manufacturers' instructions and appropriate spare parts restocked to ensure smooth running during the seasons of intense activities. Other maintenance should take place during the 'off season'.

The opportunity should be taken in low season to train new technical staff on the site and, if appropriate, at specialist training centres. Manufacturers or suppliers of seed-processing equipment may be in a position to provide training as part of their customer service. Health and safety requirements must be made clear and respected by employees and employers.

Buildings to house the machines and accommodate associated seed-processing activities should be constructed, so that they remain dry and their foundations are capable of withstanding the weight and vibration of machines. The floors and walls should have a smooth finish. Adequate provision should be made for dust extraction, clean air supply and fireproofing. Buildings should, as far as possible, be rodent- and bird-proof.

Rules of Seed Processing

- The processing should take place on a smooth clean floor, or similar conditions if outside.
- Each machine should be adequately cleaned between seed lots, using a vacuum cleaner. Removable parts must also be cleaned, especially parts where seed and debris are likely to be caught up.
- All machines should be maintained and serviced according to the manufacturers' instructions.
- Strict regards must be given to health and safety requirements.

Seed Grading

Modern commercial vegetable production systems are based on crop uniformity, a primary contribution to this is the precision drilling of seeds. Precision drills generally rely on graded seeds; the seed suppliers have therefore responded to the demand for seed of specific grades by defining the grades and offering them in their catalogues. Seed grading is often done prior to pelleting or coating.

Seed Drying

The natural moisture content of seeds is gradually reduced as they develop and ripen. By the time it is separated from the mother plant, the seed's moisture content is below 50% and therefore the moisture content is in equilibrium with the ambient or storage atmosphere.

It is frequently necessary to dry seeds when they first arrive at the processing plant from the field, after extraction from fruit or possibly after processing but before storage and packaging. The artificial drying of unthreshed seed crops was discussed by Sparenberg (1963), who reported that the depth of pea or bean material to be dried by air at 350 m³/m² of floor should not exceed 2 m, but for radish it could be up to 4 m when the unthreshed moisture content was at 30–40%. This work demonstrated and emphasized the importance of not having an excessive depth of plant material during pre-threshing drying.

Artificial drying systems can be classified into two main groups, i.e. batch and continuous systems.

Batch drying systems

These include horizontal tray, vertical single layer, vertical double layer and cylindrical bin with central duct. The horizontal tray type is especially useful for dealing with different batches of seed simultaneously.

Rotary paddle hot air batch drier
The rotary paddle driers are used for the drying of tomato, pepper, aubergine and cucurbit seeds which have been extracted from fruit by a wet extraction process.

The wet seed is placed over a finely perforated surface and dried by a current of hot air which blows through the perforations in the bed of seed from below. As the hot air passes between the seed, the moisture is removed to the outside; at the same time a rotary paddle progresses along the top of the apparatus, turning the seed in the bed. The bed of seed can be up to approximately 5 cm deep.

Advantages of the rotary paddle driers for seeds immediately after extraction from wet fruit are that they can be used for small seed lots and the temperature control is better than other batch driers. This is important when the seed is relatively wet at the start of the drying process.

For the wet-extracted solanaceous and cucurbit seeds, the air temperature is controlled between 37–40°C at the start, but when the moisture content begins to lower, the air temperature is reduced to 32–35°C.

The total time for drying a batch of wet-extracted seed depends on the crop species, but is approximately 8h for aubergine and pepper, 10h for cucumbers and tomatoes and from 12 to 16 h for the larger-seeded cucurbits such as squashes, depending on the cultivar's seed size.

When seed drying is thought to be complete, the hot air supply and the rotary paddle are stopped. Samples of seed are taken for a moisture content determination. If satisfactory, the seed is then immediately canned or taken for storage, but if necessary, it is dried further and the moisture determination repeated. Experienced workers can usually judge when the seed's moisture level has been reduced to the required level, but checks should always be made with a laboratory moisture-determining instrument.

Solar seed drier

The solar drier is an energy saving batch drier (Fig. 4.4). It consists of an inverted pyramid-shaped chamber made of black mild steel. There is a vertical chimney from the apex which exhausts to the exterior. The chamber is heated by solar radiation (see Fig. 4.4). The material to be dried is placed inside, on a tray just above ground level. As the chamber warms up the interior moist, warm air rises

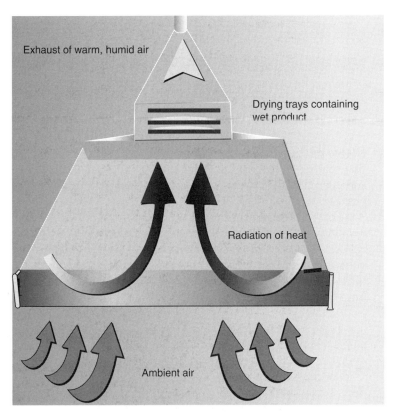

Exhaust of warm, humid air

Drying trays containing wet product

Radiation of heat

Ambient air

Fig. 4.4. Solar seed drier. Reproduced with kind permission of Alvan Blanch Ltd, Chelworth, Malmsbury, UK.

and passes up the chimney. Access to the drying chamber is through a side closable opening, so that the tray and seed material can be reached and checked for progress. The drier has a very useful role in suitable climates, especially for groups of farmers and remote locations without an easily available power source.

Continuous drying systems

These systems are suitable for relatively large quantities of seed. There are two designs based on either a vertical seed flow or a horizontal bed on a perforated belt. Kreyger (1963) described the different types of seed drying equipment used in Europe and Barre (1963) described those of importance in North America.

Drying rate and temperatures

The rate at which a seed lot can be dried artificially depends on the 'packing character' of the species and the initial moisture content of the sample. Lettuce and cucurbits are generally considered to be relatively quick driers; carrot, red beet and tomato medium driers; but legumes, brassicas, onion, leeks and sweetcorn are slow driers. Generally at 43.5 °C, 0.3% of seed moisture content is removed per hour by an air flow rate of $4 \text{m}^3/\text{m}^3$ of seed. Slow air rates are cheaper to operate but fast air flow rates prevent deterioration and allow a greater throughput with the equipment. For open storage, starchy seeds are reduced to a moisture content of 12% and oily seeds to 9%. The moisture level for seeds to be stored in sealed containers is less than these figures and is discussed in Chapter 5.

Determination of seed moisture content

The most satisfactory method of determining the moisture content of a seed lot is the method described by ISTA (2009a). This is based on an oven-drying method and the reduction in weight is calculated from the weight before drying and expressed as a percentage of moisture of the sample.

There are several types of electronic meters available and, while they are extremely useful as a guide for quick determinations, they are generally less reliable than the oven method and in addition usually require a correction factor for each crop species.

Summary of Harvesting and Processing

Machines for special purposes

- The more processing operations the greater the cost of the finished seed lot.
- Many machines have a narrow crop range and are specialized.
- Organization of throughput in processing operations.

Seed drying

- Types of drier: batch or continuous.
- Careful regard to drying rate, stages and temperatures.
- Importance of reaching and determining optimum seed moisture content for crop species.

Further Reading

Pre-harvest and harvesting conditions

Austin, R.B. (1972) Effects of environment before harvesting on viability. In: Roberts, E.H. (ed.) *Viability of Seeds*. Chapman & Hall, London and Syracuse, University Press, Syracuse, New York, pp. 114–149.

Processing

Culpin, C. (1992) *Farm Machinery*, 12th edn. Blackwell, pp. 272–315.
Delhove, G.E. and Philpott, W.L. (1983) *World List of Seed Processing Equipment*. FAO, Rome.
Gregg, B.R., Law, A.G., Virdi Sher, S. and Balis, J.S. (1970) *Seed Processing*. Mississippi State University, National Seeds Corporation and United States Agency for International Development New Delhi, India.
Vaughan, C.E., Gregg, B.R. and Delouche, J.C. (1968) *Seed Processing and Handling*. Seed Technology Laboratory, Mississippi State University, Mississippi.

5 Storage

When seed extraction and drying have been completed, it is necessary to store the seed under the best possible conditions to ensure that the maximum potential germination and other seed-quality factors are maintained. Stored seeds are the primary input of a country's vegetable cropping programme and are vital links between successive crop generations. In commerce, stored seed represents a significant proportion of the seed company's financial and genetic assets.

Reasons for Storage

The period of storage may be relatively short, perhaps even only weeks, but sometimes it is necessary to store seed for several years. There are many reasons for storing seed of individual crop species beyond the next possible sowing season. In some cases, it may be uneconomic to multiply each seed stock annually, and reasons for isolation in time have been discussed in Chapter 3. In addition, the recurrent annual cost of multiplying each cultivar offered by a seed house or seed marketing organization has to be considered. Seed yield is influenced by many factors including the multiplication rate of the species and cultivar, provenance and seasonal variations during development and pre-harvest ripening. It is not always possible to make accurate forecasts of yields; thus, satisfactory storage is a useful way of ensuring that surplus seed is kept for future use.

Storage for contingencies

An increasing number of agricultural and horticultural crop programmes in developing countries are adopting the policy of maintaining stored seed stocks for contingencies such as natural or man-made disasters. The need to safeguard seed stocks for use immediately following unpredictable drought, flooding or other disasters can result in sudden seed shortages. It can be argued that the need for seed stocks for contingencies is greater for the agricultural staple crops than for vegetables, simply on the basis of vegetable seed requirements being relatively small compared to staples such as rice, wheat and maize; also that the significantly less bulk or weight of vegetable seed required following a disaster can be supplied from elsewhere. However, in order to ensure that appropriate cultivars are immediately available, vegetable seed should be included in the seed security for food security programmes. The wider implications of the concept of reserve seed stocks against major disasters were discussed by Kelly and George (1998a). The FAO World List of Seed Sources (Sgaravatti and Beaney, 1996) has a seed security aspect in that it can be used to identify sources of potentially useful cultivars from appropriate agro-ecological areas.

Storage for fluctuating demand and seasonal differences

Farmers' and growers' demands for seed of specific crops can depend to some extent on the general market economy, market trends in crop demand or market successes or failures in previous seasons. Satisfactory seed storage will assist the seed industry to buffer these variations in requirements. In some instances, specialist seed producers are in a different area of the world from the consumer and stocks may arrive at their destination out of season. The viability of seeds can be adversely affected during the period of transit and awaiting sowing, if they are not stored properly. It is also important to ensure that stored seed is protected from hazards such as insects, fungi, fire and pilfering.

Storage of valuable stocks by seed organizations

All seed stocks have a value which is related to the level of genetic purity and multiplication rate. Breeders' and basic seed stocks are normally stored under as ideal conditions as possible to reduce the frequency of their multiplication; this makes the maximum use of each seed stock and keeps the number of multiplications to a minimum.

Storage of germplasm

The storage of seed as germplasm, a potentially valuable genetic resource, calls for very long-term storage of relatively small seed lots. Germplasm centres, usually referred to as gene banks, use specialized storage methods which have been described by Cromarty et al. (1982). The history and work of the Genetic Resources Unit (GRU), University of Warwick, have been described by Astley (2007) who also discusses the collections held, the conservation technology and future international collaboration. According to Astley, the best method for the long-term maintenence of small-seeded vegetable germplasm is low-temperature storage of seeds which have been dried to within 1% of 5% moisture content. The seeds are then packed in laminated foil, hermetically sealed and stored at −20 °C. At the time of packing, individual pouches are computer labelled. Following laboratory extrapolation, it is anticipated that orthodox seeds (i.e. not recalcitrant) will store for decades. The operating standards for conservation of orthodox seeds in genebanks have been described in a FAO publication (Anon., 1994).

Natural Longevity

The longevity of seed is primarily dependent on its inherent keeping quality, which varies with species. Some species, e.g. onion (*Allium cepa*), and leek (*Allium porrum*), are relatively short-lived. Sweetcorn (*Zea mays*), and the larger-seeded *Phaseolus* spp. are intermediate, while seed of many genera of the *Cucurbitaceae*, e.g. vegetable marrow (*Cucurbita pepo*), are relatively long-lived in storage. In addition, some workers have reported quite significant differences of storage potential between cultivars of the same species. Differences in seed longevity at the species level have been discussed by Priestley *et al.* (1985).

The pre-storage history of a seed lot will also have a very important influence on its subsequent storage qualities. Those factors affecting storage potential include mother plant environment during plant growth and seed development.

Pre-harvest Field Factors

Micronutrient and macronutrient deficiencies during plant growth and development of a potential seed crop can have a major effect on seed storage potential in that initial germination potential is low and, although such extreme conditions do not normally prevail in commercial seed production, the levels of major nutrients have been shown to indirectly affect seed storage life.

During seed production many of the environmental effects in the field which predispose the seed to a relatively short storage life are especially severe in the tropics and subtropics, but they may also occur in temperate regions. It is probably those arid regions that have adequate control over irrigation frequency and quantity which have the best pre-harvest quality control in relation to potential storage.

Field deterioration of seed prior to harvest is frequently associated with environmental factors such as high relative humidity, excessive rainfall and high temperature; combinations of some or all of these factors result in a reduced potential storage life of seeds. Some of the deleterious effects can be successfully counteracted by operations such as mechanical harvesting and seed drying. The evidence for the effects of these operations on individual crops is discussed in other sections. The effects of environment before harvesting on viability were reviewed by Austin (1972).

Pre-storage Mechanical Factors

Mechanical damage to seed during operations such as harvesting, processing or drying can reduce the potential storage life because the damaged seeds lose vigour sooner than undamaged ones. In addition, damaged seeds are more vulnerable to storage pests and pathogens. There are various forms of mechanical damage to seed which can occur before storage, one of which is threshing injury, e.g. cotyledon cracking of *Phaseolus vulgaris* (dwarf or French bean), and seedcoat abrasion of many species especially resulting from incorrect cylinder speeds. Damage can also occur during processing, if the seed is immature; the other extreme is over-drying, which produces brittle seeds that are predisposed to cracking or breaking.

The Importance of Seed Quality at Start of Storage

The different components of seed quality are discussed in Chapter 6. Those which are of direct influence on the storage potential of a given seed lot are viability and vigour. In this respect, the storage potential refers to the stage of deterioration as expressed by seed vigour. The seed lots with the higher vigour will be more able to withstand the various stresses derived from the storage environment throughout the entire duration of storage.

Definition of Storage Period

The storage period may only be relatively short, perhaps even only weeks, but it is also possible that seed lots may have to be stored for several years. The storage period of seed should be defined as the total time from seed maturity through to sowing, and although a seed lot can undergo several operations such as cleaning and packaging, or a waiting period while held in stock by a retailer or farmer before being sold or sown; it is important that this total time from maturity through to sowing be regarded as the storage period, because it is throughout this entire period that seed is subject to adverse influences of its environment. The stages comprising the total possible storage duration of a seed lot are comprised of some, or all of the following:

1. Post maturity drying.
2. Seed extraction or threshing (outside or under cover).
3. Seed cleaning (outside or in buildings).
4. Holding in seed store or warehouse.
5. Packaging (usually in a purpose-built structure).
6. Transit and distribution (all modes of transport and handling stages).
7. Marketing (wholesale and retail centres).
8. Point of sale (shop, bazaar, garden centre or farm supplier).
9. On-farm storage (from receipt of seed at the farm, or garden, to sowing).

Effects of Storage Environment on Seed

The two most important environmental factors which can affect seed quality during storage are temperature and relative humidity. In practice, the ambient relative humidity plays a major role: first, because the seed's moisture level is a function of relative humidity; and second, the incidence of storage fungi and insect pests is largely influenced by the relative humidity of the microclimate within the seed mass.

Moisture content of seed

The seed most suited for storage has a moisture content not greater than 10% of seed weight. A seed's metabolic rate is extremely low, often undetectable. When in this state, it is hydrophylic and is capable of taking in water, even from the water vapour of the atmosphere. Thus, however much care has been taken to lower its moisture content by drying before storage, it will quickly take up water again when stored in ambient conditions.

Combined temperature and relative humidity

The loss of seed viability is slower at lower temperatures than at relatively high ones. In practice, it is the combined effect of temperature and relative humidity which reduces potential viability or longevity of seed throughout its 'storage' life. Many areas of the world have periods of fluctuating temperatures coupled with periods of high relative humidity, the combined affect of which influences the rate of seed deterioration during short periods of uncontrolled storage. Harrington (1963) provided two useful 'rules of thumb' for relative humidity and temperature:

1. For each 1% reduction in moisture content, the storage life of seed is doubled.
2. For each 5 °C lowering of the storage temperature, the storage life of the seed is doubled.

The relationship between storage temperature, seed moisture content and the viability period of the seed has been discussed by Roberts (1972). The construction of viability nomographs for use as a guide to the prediction of viability under hermatic storage and to a limited extent under open-storage conditions have been described by Roberts and Roberts (1972). The storage, germination and quality of vegetable seed have been reviewed by Taylor (1997).

The extent to which it is decided to control both temperature and relative humidity will depend on the value of the seed and the climate of the storage location. In some arid areas of the world, the seed's potential germination is not reduced significantly during short periods of uncontrolled storage. This is because the natural drying of seed is satisfactory following seed maturity in the field and there is a low relative humidity during the subsequent storage period.

Even relatively short-lived seed of crops such as onion can be stored with little reduction in its germination capacity between seasons in many arid areas.

The survival potential of seed at a range of temperatures and moisture contents has been quantified by Ellis and Roberts (1981).

Subsidiary effects of relative humidity

When the seed moisture is between 40 and 60%, germination will occur, resulting in the death of the seed's embryo if it occurs during storage. In addition to the direct effect of the seed's storage life, high moisture content has indirect effects, especially on the microflora and microfauna of the seed store environment. Most storage insect pests' activities, including reproduction, are stimulated with seed moisture above about 8%. Furthermore, the growth of fungi on seed will commence when the seed moisture content exceeds 12%. At seed moisture content above 18–20%, local heating can occur as a result of total metabolic heat of the biomass; this in turn can be responsible for reducing viability or even spontaneous combustion.

Storage Environment

The storage environment can be influenced in two main ways:

- The building (or structure) in which the seed is stored.
- Its environment (i.e. temperature and relative humidity).

It is important that the seed is prepared for storage in a purpose-built structure as quickly as possible after harvesting and a system organized, so that the seed remains in the storage environment for as long as is practical prior to distribution for sowing.

Construction of seed stores

Seed stores should be designed to maximize security, minimize fire risk, exclude birds and rodents, and keep the entry of insects and microorganisms to a minimum.

The ideal building should have a raised and smooth-finished reinforced concrete ground floor with a rodent-proof lip. Entry can be by removable ramp and the raised floor designed to match up with the loading level or height of delivery vehicles. The roof should be pitched and overhung to offer the best possible runoff of storm water and provide shade and extra protection for the ventilator openings. There should also be a pitched canopy roof over the entrance chamber.

The extent to which rodent proofing, measures to deal with excessive storm water (including surface runoff water from higher levels) and extremely high temperatures are included as design features will depend on the location, topography and prevailing conditions throughout the year.

A double entrance door system incorporating an entrance lobby, or ante-chamber, should be included in the design, and ideally there should be no other openings or windows except those connected to environmental control systems.

The walls and ceiling should be constructed of smooth-finished stone, mortar and concrete which are lined with a moisture barrier of tar, aluminium foil or polythene. Wood should either not be used or kept to a minimum, as in time it can be attacked by rodents and other pests and present maintenance problems. The final finish of the interior walls should be of a material which will protect the insulation from damage made by trolley, pallets or other handling, transport or storage devices. Any ventilation openings which are incorporated must be efficiently screened to exclude insects. The overall interior finish must be smooth with electrical conduit channels and other cracks smoothly sealed.

External finishes, especially the roof, should aim to minimize absorption of solar heat and exclude water. Particular attention to these specifications and details are necessary in parts of the world subject to high temperatures and rainfall. Seed stores should be provided with an adequate lighting system but with a low heat emission. There should be a regular cleaning and maintenance programme to ensure that all the design features of the structure are upheld during the life of the store.

Additional features of conditioned seed stores

The design and materials used in the seed store's construction should minimize the absorption of solar radiation and act as an effective vapour barrier. In some parts of the world, these are the only features necessary to ensure a satisfactory storage life of seed. However, in many other areas further control of the storage environment must be achieved by air conditioning. This becomes a major requirement where the climate is such that the design of the seed store is insufficient to ensure satisfactory potential life for the seed.

The actual design and structure of conditioned stores follow the specifications outlined above. But where conditioning is required, it is extremely important to ensure that the best use is made of insulating materials in order to achieve the maximum possible efficiency from temperature controlling equipment and value from operating costs. This factor is continuing to increase in importance with the progressive rises in energy costs and the need for conservation of fossil fuels and protection of the environment.

Vapour proofing of the structure is also important and is achieved by building a continuous polythene film sealed with bitumen or other suitable sealant into all floor, wall and ceiling areas. This vapour barrier forms a seal completely enveloping the store and is installed on the 'wet' or exterior side of the heat insulation barrier. All doors must also be fitted with gaskets and there should be an air lock at the entrance.

An adequate power supply for operating the apparatus which is to control the store's environment is necessary. Control appliances should be positioned so that the heat they emit is exhausted to the store's exterior. An important consideration is positioning of air exhausts; hot air should be emitted from the store just below the roof line, while moist air should be expelled from near ground level if conditioned separately.

Temperature control

Storage temperature can be reduced by ventilation and refrigeration in addition to insulation and structural features. Complete temperature control is expensive in any part of the world and is unlikely to be used in commerce for seeds, regardless of their value. Total temperature control by refrigeration is, however, used in the long-term storage of germplasm collections and breeders' material and was outlined above, under 'Storage of germplasm'. Work by Ellis *et al.* (1996) has indicated that storage at −20 °C rather than 20 °C is beneficial to seed survival, and that hermetic storage at 20 °C of seeds first dried at 20 °C to moisture contents in equilibrium with about 10% RH provides greater longevity than 5.5–6.8% moisture content in the five species studied (carrot, groundnut, lettuce, oilseed rape and onion).

Temperature control during the storage of 'short-term' and 'carry over' seed stocks is usually achieved by ventilation in conjunction with refrigeration, the degree of which will depend on the outside temperature.

Ventilation

The ventilation of seed stores should be considered in conjunction with the relative humidity of the ambient air, because it would be more harmful to the seed to lower its storage temperature, if the result is to increase the seed's moisture content. Conversely, ventilation can be used to lower the storage temperature and the seed's moisture content when the relative humidity of the outside air is relatively low.

Storage engineers can design systems controlled either manually or automatically which operate ventilating fans according to the temperature and relative humidity of the outside air. Ventilation also enables a gentle air flow to be passed through bulk lots of stored seed as and when required, thus ensuring that hot spots do not develop which would endanger the stored material.

Another form of localized heating sometimes coupled with moisture migration can occur as a direct result of convection currents. These are most likely to occur in unattended stores with relatively poor insulation and are caused by drier warm air moving from a warm spot to a cooler one within the store. On cooling, moisture is condensed which is subsequently absorbed by the drier seeds.

Refrigeration

The use of refrigeration in controlling seed store temperatures is generally confined to long-term storage of high-value material, for example, germplasm collections and breeders' stocks (Astey, 2007; Cromarty *et al.*, 1982). However, refrigeration is also useful in the tropics for valuable categories of other seed stocks.

Extra care and attention must be given to thermal insulation and structure of the store when refrigeration is to be included in the control systems.

There are four sources of heat within a seed store which the refrigeration has to cope with; these are derived from:

- Leakage from outside (despite insulation).
- Field heat of seed and associated materials.
- Respiration of seed in the store.
- Incidental heat derived from lighting, other equipment, workers and external heat which enters when doors are opened.

The cooling coils of a refrigeration unit should be situated within the storage area but the compressor must be sited so that its heat is given off to the exterior of the storage room. The relative humidity of the air is reduced during the refrigeration process. The moisture condenses on the cooling coils which have to be defrosted at intervals. Although this reduction in relative humidity is an advantage, in practice, the store's relative humidity is inversely proportional to its temperature and at temperatures below about 13.0 °C, the relative humidity is too high for open storage.

Dehumidification
An alternative system to refrigeration of seed stores is the use of a suitable chemical desiccant in a dehumidifier.

There are two types of chemical dehumidifiers generally used in seed stores: the bed and the revolving drum. In each system, the apparatus will add to the interior heat load if not carefully sited and it is therefore important that dehumidifiers be placed in the structure so that their heat is given off to the outside of the store. Silica gel, which can absorb up to 40% of its own dry weight in water, is usually used for seed store systems. In the bed system, the silica gel is dried up to about 175 °C to drive off all the absorbed moisture. After cooling, air from the storage area is blown through the dried silica gel bed. When the silica gel is again in moisture equilibrium with the air, it is reheated to dry it before further reuse. Some bed systems use two beds per unit, in which case one bed is dehumidifying the store's atmosphere while the other is being dried to reactivate it. The operation of alternating beds is normally controlled by a time clock.

The revolving drum system has a desiccant bed which is divided between two air streams. Different sections are dried or used for absorption as the bed rotates. The revolving drum systems are capable of removing more moisture from a given air flow than the bed systems.

The choice of system will depend very much on local conditions and storage requirements, so in all cases qualified and experienced environmental control engineers should be consulted.

Storage in Vapour-proof Containers

There have been major developments which have led to the storage of seed in sealed moisture-proof containers. Most of the original research and development work with this technique has been done with seed of vegetable species because of the relatively small seed lots required by individual farmers and, with the exception of the larger-seeded vegetable species in *Leguminosae*, the proportionately high value of vegetable seed per unit volume is compared with cereals and most other crop groups.

The principle is that seed lots are dried to a moisture level slightly lower than they would be prior to normal open storage, and are then sealed in metal cans, packets or other suitable moisture-proof containers. As a result of this containerization or packaging, each seed lot is in its own environment and may be stored at ambient temperature and relative humidity for 1 or 2 years, or even longer, with little or no deleterious effect on germination. In fact, provided that careful attention has been given to important details such as drying, and the determination of moisture content, and appreciating that different seed lots can differ in their composition, it is possible that the potential storage life of a seed lot at 8% moisture content can be doubled if 1% more moisture is removed before sealing.

Seed moisture before vapour-proof packaging

It is of fundamental importance that seed moisture content is reduced to a satisfactory and safe level before the seed is sealed in vapour- or moisture-proof containers. This is generally 2–3% lower than for other forms of storage or packaging in non-moisture-proof containers. The reason for this is that the atmosphere within a moisture or vapour-proof container will equilibrate to the moisture level which is in the contained seeds. This results in a long-term relative humidity which is too high for safe storage. For example, if sweetcorn seed with 13% moisture content is sealed in a moisture-proof container, the enclosed storage atmosphere will equilibrate at a relative humidity of approximately 65%, which is too high for safe seed storage. In addition, some storage pathogens will become active within the sealed container and the seeds' respiration rate will be relatively high.

Safe maximum moisture content in open storage
(i.e. *not* in vapour-proof packets)

The safe moisture content, in open storage, for the majority of small-seeded vegetable species is generally 8%; for leguminous species (e.g. peas and beans) 10%; for *Zea* sp. (sweetcorn and maize) 13%.

Safe moisture content in vapour-proof packages
and containers

The safe moisture content for vegetable seeds in vapour-proof packages and containers is generally considered to be 5–6%; details of this type of container follow below.

The storage of vegetable seed in vapour-proof containers

The storage and distribution of vegetable seed with low moisture levels in vapour-proof containers have been widely adopted throughout the world. There is a range of containers used for this purpose which are discussed in Chapter 6. The most important use of vapour-proof containers or packages is probably the shipping and marketing of high-value seed, such as hybrid seed to the humid tropics; but vapour-proof containers have also had a major influence on seed longevity, storage and distribution in temperate regions and have become the norm for packaging and distribution of vegetable seed for commercial and private growers.

Many seed companies in temperate regions store their processed commercial and stock seed in a section of their warehouse or store which has a controlled atmosphere; the operation of filling and sealing moisture-proof containers is carried out within the store. In arid areas, seed producers containerize the seed from the open store, usually immediately after drying, but first checking the moisture content of each seed lot and further drying if necessary.

A simple and cheap method for the storage of small quantities of onion seed without refrigeration, used in Zimbabwe, has been described by Currah and Msika (1994); the system uses airtight jars, with or without dried rice as a desiccant.

Seed Store Management

The level of hygiene within the store will have a long-term effect on seed quality and longevity. Only seed which has been through the final stages of processing should be taken into the store. All other materials should be excluded. In practice, it is sometimes tempting to use seed stores for short-term retention of other plant materials such as dormant bulbs or selected fruit waiting for seed extraction, but this misuse of the seed store can lead to the introduction of storage pests and pathogens. Plant materials other than seeds are very likely to add to the moisture content of the storage atmosphere. Other sources of water and moisture should also be excluded in order to discourage rats and other rodents.

A comprehensive programme for rodent prevention should be organized from the outset, rather than waiting for any control measures to become necessary later. The possibility of rodent infestation will depend on the method of containerization within the store as well as on the location and local conditions.

Rodent prevention and control programmes include the use of rodenticides such as the blood anticoagulants or other proprietary poisons. The material should be used in accordance with the manufacturer's instructions and current national legislation relating to use of poisons and safety. Other methods to deter rodents from stores and warehouses include the installation of sonic sound systems which are undetectable by workers. The repetitions of sound recordings of birds of prey are also used as deterrents.

Seed stores should not be used for storage or shelter of machinery, apparatus or any other materials not directly involved with the stored seed. Additional apparatus and materials make it difficult or even impossible to maintain a high degree of cleanliness. All surfaces should be kept clean and floors should be cleaned with vacuum cleaners in preference to brooms in order to minimize the build-up of dust. All waste materials should be removed from the store as soon as they are accumulated and disposed of by incinerating safely, as far away from the store as is practical.

Seed stores have a relatively high fire risk due to the dry nature of seed and the possibility of dust in the atmosphere. It is therefore vital that adequate fire prevention measures are formulated and all staff are familiar with them.

A system for entering, locating and retrieving seed lots should be adopted. The system should take into account the need for sufficient space between seed lots for access and air circulation. Small seed lots should be on suitable shelving and large quantities in bags or sacks should be neatly stacked on pallets.

The most sophisticated seed stores belonging to the larger seed companies have a computerized and fully automated retrieval system. But the main criteria for any system, regardless of the size of operation or level of sophistication, is that all bags, cans or any other form of container must have a label inside and be clearly identified on the outside. Labels must be firmly attached; adhesive labels on tins or other containers should not peel off when subjected to the storage environment. The labels should be written in accordance with the inventory system adopted, which should maintain a record of each seed lot's year and other details of provenance, designation, samples tested, quantities removed and balance remaining. This information will then always tally with stock books and other records.

The structure and fabric of the buildings should be regularly inspected and any deterioration or damage must be restored immediately by competent tradesmen.

Summary

Definition of storage period

- The overall storage duration starts at harvest and continues until sowing.
- The storage period may be only weeks in some cases, or seed lots may have to be stored for several years.

Harrington's rule of thumb for seed storage

1. For each 1% reduction in moisture content, the storage life of seed is doubled.
2. For each 5 °C lowering of the storage temperature, the storage life of the seed is doubled.

Reasons for storage

- Contingencies, e.g. poor seed harvest, natural and man-made disasters.
- Fluctuating demand and need to carry seed over to the next season.
- Maintenance of maximum seed vigour and potential germination.
- Maintenance of valuable seed classes, germplasm and breeding material.

Seed store construction and management

- Pay attention to optimum structural requirements so as to maximize storage environment requirements.
- Choose store site, considering local year-round weather and environmental conditions.
- Correctly choose structural materials and final surface finishes.
- Attend to potential fire, vermin and safety hazards.
- Instil seed store discipline, including exclusion of all non-seed store materials and items.
- *Do not* store chemical, pesticide or fertilizer materials in the seed store.

Remember the following:

- Most storage insect pests' activities, including their reproduction, are stimulated with seed moisture above about 8%.
- The growth of fungi on seed will commence when the seed moisture content exceeds 12%.
- At a seed moisture content above 18–20%, local heating can occur as a result of total metabolic heat of the biomass; this in turn can be responsible for reducing viability or even spontaneous combustion.

Further Reading

Astley, D. (2007) Banking on genes. *The Horticulturist* 16(2), 2–5.

Copeland, L. and McDonald, M. (1995) *Principles of Seed Science and Technology*, 3rd edn. Chapman & Hall.

Cromarty, A.S., Ellis, R.H. and Roberts, E.H. (1982) *The Design of Seed Storage Facilities for Genetic Conservation*. International Board for Plant Genetic Resources, Rome.

Justice, J.O. and Bass, L.N. (1978) *Principles and Practices of Seed Storage*. Agriculture Handbook Number 505, USDA, Washington, DC.

6 Seed Handling, Quality Control and Distribution

Seed Organization or Seed Company Records

These should record the successive stages from field production to assembly and distribution of the final product. In some countries it is a legal requirement to maintain records of the origin and identity of each seed lot. Well-documented and maintained records are also essential for the future planning at farm level and for dealing with problems relating to seed quality, such as growing-on tests, admixture or cultivar purity (i.e. genetic quality), and seed-borne pathogens.

Records during planning and production

These should include the following information: rotation and cropping frequency of individual sites; applications of pesticides throughout the production of the crop (including pre-sowing or pre-planting treatments) partial sterilization of substrate and soil fumigants; cultural details, e.g. sowing rates and dates of sowing, planting and other operations, applications of fertilizers (organic and inorganic) during preparation and growth of the crop; roguing stages and dates with identity of responsible staff, details of off-types and percentage removed or selected plants retained; observations on effective isolation and seed yield.

In the case of organically produced seed the records should indicate that the site used has been managed organically (i.e. according to prevailing specifications for the production of 'organic seed') for at least 2 years, or the minimum period stipulated by the monitoring agency, before it is eligible for organic certification.

Records of a seed lot

These include a stock or lot number that is designated on arrival of the seed lot and against which all subsequent information on that lot is recorded. Stock numbers are allocated from a stock book and may be formed from a coded system of digits and letters including stock number of parent material and information such as year of multiplication, seed generation or category, standard of processing and processor's identity.

All seed lots received are entered into the stock book including those produced by the company or organization, contractors or other seed companies. The records should include: year of multiplication, provenance, pre-cleaning and processing operations and quantity of seed after the final processing operation.

Additional information recorded includes weight of a sample at a given moisture content (10 g is normally an appropriate weight for vegetable seeds) and the results of germination, purity and growing-on tests. Further data such as results of annual germination tests are added during the time that the seed lot is kept by the company or organization, including any seed treatments during storage or prior to dispatch.

Security of Stored Seed Stocks

Factors affecting longevity of stored seeds were discussed in Chapter 5 but the monetary value of seed stocks is again emphasized here. Particular attention must be given to overall security, fire and attack by vermin or insects. The identity of seed lots must not be lost or mistaken. It is customary to have duplicate labels on both the outside and inside of containers, and all staff must be trained to check both of these when handling the seed lot. It is advisable that only authorized members of staff be allowed to withdraw or deposit seed in the store.

Seed Quality

Each of the following contributes to the overall quality of a seed lot:

- Mother plant environment, including choice and suitability of site for multiplication.
- Previous cropping history of the site.
- Favourable and/or unfavourable field events from sowing of the crop throughout the field production phase, ripening in the field and harvesting conditions.
- Harvesting, drying, threshing, processing, storage and rate of deterioration during storage.

Monitoring Seed Quality

The reasons for assessing the quality of a seed lot are numerous and include determining the value, purity and moisture content. The different aspects of quality as defined in the section above are used to evaluate the sowing value of a specific seed lot, i.e. its true value to farmers and growers. The criteria may also be used to determine the commercial value, i.e. the price paid to the contractor who produced the seeds or vendor in a subsequent commercial transaction, and ultimately to protect the final purchaser of the seed who uses it to produce a vegetable crop. Different evaluations may be done in the pathway of the seeds from original producer through to the grower who sows them, but each test or evaluation generally follows an internationally agreed and accepted procedure, and this is essential for evaluations subject to legislation.

The criteria for the assessment of seed quality are referred to as *attributes* while the factors, or causes, that may predispose the seed lot to any of the attributes are known as *determinates*. It should be noted that there can be several determinants of both pre-harvest and postharvest origin that may affect an attribute, these are summarized at the end of the chapter.

The seed quality attributes are:

- Germination and vigour.
- Genetic purity (trueness to type).
- Mechanical purity.
- Seed health.
- Moisture content.

Genetic quality: this may be referred to as 'trueness to type', cultivar purity or varietal purity. The strict control of seed generations or categories coupled with implementation and monitoring of seed certification procedures (or other monitoring schemes where they apply) provide ways of assisting the authenticity of a seed lot.

Physical purity: the range of components of a seed lot can include seed of the stated species (i.e. pure seed), seed of other species (i.e. including other crop species and weeds) and contaminants. The contaminants may be derived from a range of sources including materials from seeds, parts of plants, living organisms (but not of plant origin) and other materials such as soil. The pure seed definitions of a wide range of crop species, including vegetables, have been published by ISTA (2009a).

Seed health: this may be described as the extent to which seed-borne pathogens and/or pests are present. ISTA has published a list of seed-borne diseases with annotations regarding bibliographies and treatments (Richardson, 1990).

Viability: this can be referred to as the potential germination and subsequent production of a seedling of the stated cultivar. Although the viability of an individual seed can be determined, it is more usual to refer to the germination potential of a seed lot. In this context, a seed lot is taken to be from a specific population, i.e. it is composed of a homogeneous population derived from the same stock, from the same production location at the same time and therefore having the same reference number; it is also assumed that all seeds in the seed lot have been treated in the same way regarding processing and storage. It is essential that methods used for determining the germination of a sample taken from a seed lot are both repeatable and reproducible; this is especially important from the aspects of legislation and international trade.

The seed testing procedures described by ISTA (2009a) provide an international standard. The seed testing rules set by AOSA (2008) are used in the USA and Canada, although there is very close cooperation between AOSA and ISTA.

Vigour: defined by ISTA (1995a) as 'the sum total of those properties of the seed which determine the level of activity and performance of the seed or seed lot during germination and seedling emergence'. The tests that are recognized are described or listed by ISTA (1995b) and AOSA (2008).

Moisture content: this is the percentage of moisture of the seed lot. Although there are quick electronic systems in use, moisture content is usually determined by an oven drying method as specified by ISTA (2009a).

The International Seed Testing Association (ISTA)

The main purpose of ISTA is the promotion of uniformity in seed testing procedures. To achieve this it produces the International Rules for Seed Testing that embrace the advice of its various technical committees. The rules prescribe principles and definitions while the 'Annexes' describe methodology; both of these are published as a supplement to the Association's journal *Seed Science and Technology*.

The Association's secondary purpose includes the promotion of research in the subject areas of seed science and technology and the encouragement of cultivar certification. These objectives are achieved by organizing and participating in conferences and training courses.

ISTA has developed a system for the quality assurance of seed testing through its system of accredited laboratories. This embraces international standards and audit visits to ensure that requirements such as staff training, operating procedures, test records and laboratory independence.

In 2009, there were more than 70 member countries and over 1500 subscribers. The Association produces an information pamphlet, publishes the *ISTA News Bulletin* three times a year and organizes a triennial World Seed Conference.

Role of ISTA in evaluating seed lots for GM material

- Produce additional seed testing rules for the detection, identification and quantification of GMO material in conventional seed lots.
- Organize proficiency testing for the detection of GMO material in conventional seed lots.
- Set up a platform for the exchange of information between seed testing laboratories.
- Additional objectives of the task force include identification of stacked genes and publication of performance test results and the availability of documentation relating to testing for specific traits.

The work and future programmes of the task force have been described by Haldeman (2008).

The Association of Official Seed Analysts (AOSA)

In North America, the Association of Official Seed Analysts (AOSA, 2008) plays a similar role to ISTA. The Association's constitution and by-laws were published in 1993 (AOSA, 1993). The history, legal framework, technical achievements and possible future developments of AOSA have been reviewed by Bradnock (1998b). AOSA publishes a journal *Seed Technology and Seed Technology News*. AOSA and ISTA work in close collaboration on technical matters.

Seed legislation

The general objectives of legislation related to seed are to ensure that seed that is offered for sale to farmers conforms to minimum required standards of quality, i.e. cultivar purity, health, vigorous germination and freedom from adulterating materials including weed or other crop seeds. Coincidental to these objectives are the protection of the vegetable grower from unscrupulous vendors. In countries where the seed industry is less developed, the latter point may be more important but once a seed industry is established, it is the monitoring of the different facets of seed quality that gain importance, especially when certification schemes are introduced. This subject is discussed further by Kelly (1994), national and international examples are detailed in Kelly and George (1998b).

Packaging

A wide range of materials and types of containers is used for the marketing and distribution of seed. The range includes:

- Burlap (a course canvas), cotton, paper (bleached sulfate and bleached Kraft papers are both used).
- Polythene or other suitable plastics.
- Hessian (bags made of this are frequently known as gunny bags).
- Aluminium foil, tinfoil and tinplate.

Many of the vapour-proof packets are made from laminations of two or more materials, such as aluminium foil and polythene; aluminium foil, paper and polythene or laminations of paper and asphalt. Glass containers are normally only used for laboratory reference collections or training purposes to demonstrate seed; they are not suitable for seed distribution or marketing.

The choice of material and size or capacity depends to some extent on the level of development of the seed industry; packaging machine capability;

the destination; type of market; mode of distribution; protection required from hazards such as inclement weather, high relative humidity, rodents, insects, pathogens; and the amount of handling during transit.

Size of containers

The quantity of seeds and size of container or packet are usually adjusted according to the intended market, for instance, relatively small quantities in packets for the amateur or private gardeners. The exact quantity may be adjusted to potential germination, purity and the selling price per packet.

For professional and commercial growers the quantity is usually in units of a gram or kilogram, although for some market outlets the unit of weight remains as pounds and ounces. Another possibility is for the contents of a seed packet to be based on a specified (or predetermined) number of seeds. Figure 6.1 illustrates a range of seed package sizes and seed quantities prepared by the government seed service available to vegetable growers in Malaysia.

Modern seed companies have very sophisticated packaging lines attached to their warehouses that are capable of automatically filling containers by machines, delivering predetermined quantities of seeds into packets and sealing and applying labels to the completed vapour-proof packet. The packaging may take place in a controlled environment similar to the storage conditions.

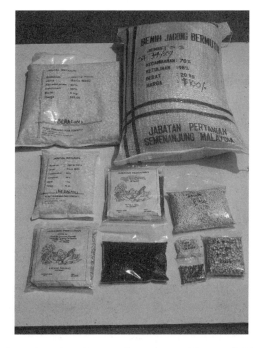

Fig. 6.1. Range of seed-packet sizes and seed quantities available from the Malaysian Government Seed Service.

Labelling

The label on a seed package is the normal way of immediately identifying the contents. Labelling systems used in seed stores and for warehousing usually conform to the system devised for the company's stock book.

Labelling for the market usually has to meet the legal requirements of the country in which the seeds are to be marketed, and includes information conforming to legal requirements and certification schemes or any other prevailing quality assurance system. Seed companies also provide additional information according to the market outlet (i.e. private or commercial growers).

Information on labels

The following information may therefore be provided on labels:

- Name of the seed company and its trademark or logo.
- Species and cultivar name.
- Year of packaging and/or legal 'sell' or 'use by' date.
- Germination, purity and statement regarding noxious weed seeds.
- Seed treatments (e.g. Thiram or hot water treatment of celery seeds).
- Batch number or stock code (this will give the seed company immediate reference to the seed lot in any subsequent enquiry or complaint).
- Class of seed (internationally agreed colour codes are used for different seed classes).

Seed supplied to commercial growers is sometimes presented in a package or container suitable for a specified area of land. Packages for the amateur or private market usually have a pictorial label that gives an indication of the species' and cultivar's characters. In addition, basic cultural information may be added such as recommended sowing date, husbandry methods, including frequency of successional sowings. Where seed companies export to numerous different countries, some of this information is presented diagrammatically or in several languages.

There is an increasing tendency for the more advanced seed companies to supply seed by count; this is of importance for growers using precision drills and who aim to have a predetermined plant population per unit area of land.

Promotion

A seed company's promotion of its products and the activities of governments, organizations and agencies in developing countries to focus farmers' and growers' attention on improved seeds are important aspects of marketing and adoption of new cultivars.

In addition to their commercial role, these activities play an important part in agricultural and horticultural development. Other influences or sources of information on the use of improved seeds or new cultivars are the advisory and extension services, recommended lists (including lists of cultivars found to be suitable for organic crop production), and the press, radio and television services. These sources of advice and information are often catalysts for the adoption of new crop production systems coupled with improved seeds or particular cultivars.

Demonstrations, open days and shows

There is probably no better way of illustrating the merits of specific seed stocks to farmers and growers than growing them in demonstration plots under the same ecological conditions as prevail in the vegetable production industry. Seedsmen therefore organize demonstrations of crops grown from their material. At appropriate times of year 'farm walks', 'open days' or 'cultivar demonstrations' are organized. These may be held in conjunction with other interested organizations such as extension services, processing organizations or growers' cooperatives. In developing countries cultivar demonstrations may be organized at village level by agencies and organizations (including non-governmental organizations) involved with population health and nutrition, relief programmes and education. In some countries or areas it is important to take gender issues into account, especially where vegetable crop production is traditionally the responsibility of women.

Catalogues

The majority of seed companies produce an annual catalogue and price list, normally in sufficient time for growers to place their seed orders. Cultivars are listed for each crop species, and it is customary for a popular description of the cultivar to be included with particular reference to specific season of use, morphological type and resistance to specific pathogens. Additional information may include special recommendations, e.g. success of the cultivar in trials. The catalogue number of each item is usually used when placing orders and conforms to reference numbers used in the dispatching warehouse. Seedsmen's catalogues usually contain other relevant information including conversion tables, numbers of seed per unit weight, recommended sowing rates, programmes for successive production, and may also offer optional services for some species such as seed grading, priming or pre-sowing treatments for the control of specific pathogens or pests. Catalogues or lists of organic seed of cultivars suitable for production in organic systems are produced by those companies that have entered this market. Other items included are statements relating to current seed laws and trading conditions.

Participatory plant breeding

There is an increasing recognition for the encouragement and establishment of linkages between agricultural research and farmers in developing countries (Eponou, 1996). The improved liaison between the end user and the researcher (including plant breeders) offers an excellent opportunity for farmers and their dependents to express their views and influence the outcome of breeding programmes and resulting cultivars.

The term 'participatory plant breeding' (PPB) is used for plant breeding programmes in which the developing lines or plant material, while still under the jurisdiction of the breeder, are evaluated on a site, or sites, on which the breeding lines are cultivated under local growing conditions. It has the objective of encouraging local growers and communities to participate in the selection process and for them to have an influence on the development of cultivars for their specific agronomic conditions and product requirements. For example plant breeders and farmers have jointly contributed to the evaluation of bean cultivars in Rwanda (Anonymous, 1995).

Thus, the application of PPB has the possibilities of embracing any, or all, of the following: growers, consumers, marketers, processors, policy makers and nutrition and health experts. Where required, specialists in food security or gender issues can also be involved in the selection process.

The essential advantages of PPB are that developing breeding lines are evaluated under the future user's conditions and that all requirements are taken into account during the selection and cultivar development process.

This system of evaluation and cultivar development is used, where appropriate, by many plant breeders in both public and private organizations. For example, the Center for Agricultural Research in Dry Areas (ICARDA) ensures that the local growers' cultural conditions and requirements are taken into consideration. PPB may also be applied by private seed companies in breeding programmes to develop cultivars for organic production.

Public relations

The public relations activities of seed companies often include a technical advisory service, and technical representatives who, in addition to their sales activities (see later), are also involved in offering advice and dealing with customer queries or complaints. The technical staff generally has a very good local or regional knowledge of the seed industry and crop production technology. Reputable sections of the seed industry are often too hastily blamed for poor seedling emergence or crop stand and the diplomatic attitude of a seed company's technical staff can sometimes be supported by results of independent laboratory or growing-on tests. Conversely, a problem occurring with the same original seed lot distributed to different growers may be resolved by reference to seed test or field plot records. There has been a general increase in the amount of printed technical information available from seed companies especially that produced for commercial vegetable growers.

The participation of seed companies at shows and exhibitions is also a way of displaying cultivars and their merits marketed by a particular seed company or organization. Some seed companies sponsor prize money at local shows for winners who have used their seeds. There has been a general decline in seed companies exhibiting at agricultural and horticultural shows in some countries due to the high cost of producing and mounting exhibits and the need to justify the cost-effectiveness. However, the advent of permanent show grounds has assisted some institutions and organizations that serve the vegetable-producing industry to organize and produce demonstrations that include the growing crop. The concept of 'grower education' in a modern vegetable production industry involves a multi-organizational approach including educational, research, extension and commercial activities; the closer the liaison between these different types of organization the better the prospects for the grower and the industry as a whole.

Price Structure

The demand for seeds in countries with an established vegetable production industry is relatively stable compared with that in developing countries in which there may be a rapid transition from total reliance on farmers' own saved seed to seeds produced by government agencies or commercial organizations. The relatively complex operations in vegetable seed production that involve factors such as biennial species, vernalization requirements, maintaining parents of F1 hybrids, producing hybrid cultivars and the skill required to maintain the required genetic quality tend to minimize growers' motivation to produce their own seeds except where economic or social conditions dictate it.

There is evidence that in countries where new seed industries have been set up, it is beneficial in the long term to sell seeds at a price level to cover all the production costs plus a profit for the vendor (Douglas, 1980). Generally, the more realistic the charge made for seeds produced by government agencies in the early stages of seed industry development, the sooner the private sector will wish to participate. There is, therefore, a better basis for a realistic seed price in the vegetable seed industry than perhaps in other areas of agriculture, especially in countries with a recently created demand for increased quantities of seeds. However, some government initiatives relating to legislative requirements (e.g. seed testing and certification schemes) will tend to increase the overall production costs.

The components of seed price can be broadly classified into two groups, i.e. direct costs and indirect costs: direct costs: these include the factors that are directly related to the quantity of seeds marketed, such as costs of stock, field production, processing, containers and transportation; and indirect costs: these include activities relating to promotion such as maintenance of trial grounds, production and distribution of the catalogues.

Distribution

The relatively high monetary value per unit weight and perishable quality of vegetable seeds makes the need for care during transportation and distribution a high priority. The deterioration of seed quality is generally minimized by the use of vapour-proof containers and packages (discussed in Chapter 5). However, factors that may contribute to deterioration of the containers, labelling system and seeds not transported in vapour-proof modules include changes in temperature and relative humidity, condensation on arrival in high temperature zones as a result of chilling during airfreight, contamination by other cargo and materials during transit, short-term storage and final distribution. It is, therefore, extremely important that seed organizations and their agents appreciate the hazards that the product is exposed during distribution and marketing.

A commercial company marketing seeds internationally usually has a marketing manager for each region. Within a region there are usually distributive agents or the company may have its own direct sales outlets.

On a national basis, companies usually market their seeds via agents, wholesale or retail outlets, or their own sales staff whose role is to promote their company's products in a specified geographical area and to obtain orders. In this latter case their sales staff may also be responsible for the supply of stock to outlets such as shops and garden centres.

Summary of Effects on Seed Quality Attributes Relating to Agronomy, Handling, Harvesting, Processing and Storage

Seed quality attributes	Pre-harvest and postharvest determinants and/or operations influencing seed quality attributes
Germination and vigour	Seed-ripening conditions on the mother plant
	Postharvest ripening
	Drying
	Processing (avoidance of mechanical damage)
	Storage including package and container environment
	Adverse and/or untimely seed treatments
Genetic purity (trueness to type)	Accurate labelling at all stages and times
	Purity of original stock and/or basic seed
	Control of ground keepers (volunteer crops)
	Isolation (in time and distance)
	Roguing efficiency, at appropriate stages
	Implementation of cultivar descriptions
Mechanical purity	Processed seed free from seeds of other crop species, weeds or parasitic plant species
	Efficient winnowing, seed cleaning and correct use of appropriate specialist seed cleaning machines, depending on contaminants present in the seed
Seed health	Stock seed free of seed-borne pests and pathogens
	Adequate control of volunteer plants in the field that are alternative hosts to pests and pathogens
	Roguing out infected mother plants as soon as identified
	Timely and efficient seed treatments
Moisture content	Harvesting conditions, timely harvesting and drying

Further Reading

Douglas, J.E. (1980) *Successful Seed Programmes*. Westview Press, Boulder, Colorado.

Gregg, B.R. (1983) Seed marketing in the tropics. *Seed Science and Technology* 11, 129–148.

Gregg, B.R., Delouche, J.C. and Bunch, H.D. (1980) Inter-relationship of the essential activities of a stable, efficient seed industry. *Seed Science and Technology* 8, 207–227.

Haldemann, C. (2008) ISTA and biotech/GM crops. In: *Seed Testing International*. ISTA News Bulletin No. 136, October 2008, 3–5.

Kelly, A.F. (1994) *Seed Planning and Policy for Agricultural Production*. Wiley, Chichester, UK.

7 *Chenopodiaceae*

The main genera in this family that are cultivated as vegetables are:

Beta vulgaris L. ssp. *esculenta*: beetroot, red beet
Beta vulgaris L. ssp. *cycla*: spinach beet, leaf beet, chard or Swisschard
Spinacea oleracea L.: spinach

There are also other species that are of local importance, especially as edible leafy vegetables. *Chenopodium quinoa* Willd., commonly known as quinoa, is grown in South America, particularly in the Andes, as a staple grain crop and to some extent for its edible leaves.

Beetroot: *Beta vulgaris* L. ssp. *esculenta*

The cultivated beetroot with its swollen root is derived from *B. vulgaris* ssp. *maritima* (L.) Thell. Sugarbeet, which is an important industrial crop, and mangold (also known as mangel-wurzel), grown as a root crop for stock feed, are also forms of *B. vulgaris*; they are both closely related to beetroot to the extent that they freely inter-pollinate.

Beetroot, which is generally cultivated for the production of its swollen roots, is a popular crop in the Middle East, Europe and North America. Processors have shown an increased interest in this crop in recent years. It is not widely grown in the humid tropics, but is increasing in popularity in the cooler tropics.

Criteria for development and selection of the modern cultivars have been based on root shape and colour, appearance of the root's transverse section, shape and colour. Root morphology includes globe-, cylindrical- and long-rooted types. Plant breeders have selected lines with resistance to early vernalization in the development of cultivars suitable for early sowing.

The 'seed' of beetroot is formed by an aggregation of flowers fused by their swollen perianths to form a multigerm fruit, which is a cluster of two to five seeds. Plant breeders working with sugarbeet have developed monogerm types in which the flowers remain unattached to each other and the fruits therefore bear only a single seed. This character has important agronomic advantages, allowing the seeds to be sown singly for precision drilling and optimum spacing of single plants, obviating the need for singling. The monogerm character has also been transferred to beetroot, but not all cultivars are available with this character; the majority still offered by the seed trade being multigerm.

Cultivar description of beetroot

There are open-pollinated and hybrid cultivars. The following is a basis for cultivar description:

- Germity: (i.e. monogerm or multigerm).
- Seedling: anthocyanin coloration of hypocotyl – absent or present.
- Leaf: attitudes and morphology of petiole and blade, degree of anthocyanin.
- Root: position in soil, shape, relative length and width, colour:
  - Transverse section, colour, absence or prominence of rings.
  - Shape of longitudinal section.
- Bolting tendency (from an early sowing), suitability for early sowing.
- Resistance to specific pathogens and pests.

(A detailed test guideline (060) is obtainable from UPOV; see Appendix 2.)

Soil pH and nutrition

Beetroot are slightly tolerant to acidic soil conditions, i.e. soils with a pH between 6.0 and 6.8 are suitable.

The general ratio of fertilizers applied during the seedbed preparation is N:P:K 2:1:2, although some growers prefer to give a lower proportion of nitrogen than this and apply 1:1:2 during preparations, with further nitrogen given as top dressings in the first season. A spring dressing of nitrogen is given to the transplanted stecklings at the rate of 50 kg/ha. Supplementary nitrogen is also applied as a top dressing before flowering in areas with a relatively high leaching rate.

Soils with a low boron status should either be avoided or a supplementary dressing of boron applied during preparations. Ideally, a boronated fertilizer is used for the main base dressing or, if not available, boron is applied as sodium tetraborate (borax) at the rate of 1.5–2 kg/ha. The main symptoms of boron deficiency in beet are black cankerous areas on the root exterior and also between the concentric rings. This interior symptom is clearly seen when affected roots are cut transversely.

Manganese deficiency occurs in some specialist production areas, e.g. in Washington state, USA, where the problem is counteracted by dressings of up to 100 kg/ha of manganese sulfate during seedbed preparation.

The main beetroot seed production areas grow the crop on the flat, but in the Middle East, where furrow irrigation systems are used, the crop is grown on ridges.

Seed production methods

There are two basic methods of seed production for beetroot; these are 'seed to seed' and 'root to seed'. The 'seed to seed' system is normally only used for the final stage of seed multiplication while the 'root to seed' system is used for the production of basic seed and also preferred in some areas for the final multiplication. This latter system allows for inspection and roguing of roots.

Root to seed
This system is in two stages. The first stage is very similar to the production of beet for market. Seed is sown during July–August in areas such as California, where growing conditions are satisfactory into the autumn. The seedlings are singled as soon as possible following emergence to allow the root to develop its characteristic shape, but this operation is not normally necessary if monogerm or rubbed seed has been precision drilled.

The roots are lifted in the autumn, selected for root characters and the desirable material stored. The lifting operation is completed before the onset of damaging frosts, and thus the timing depends on the local climate. Whatever lifting procedure is adopted, every care must be taken to avoid mechanical damage to the roots. The lifting and subsequent root handling is easier if the plants are 'topped' first by passing over the crop with a cutter bar or mower. The crown of the root must not be damaged during topping or lifting.

ROOT STORAGE   The two basic systems for storage of roots are either the use of suitable buildings or storage in the field in clamps or pits. Most beetroot seed producers have developed suitable shed systems, but field storage is still used in some areas.

Storage in buildings: the main advantage of storage in buildings is that the air temperature can usually be controlled when necessary, thus avoiding over-heating and frost damage. The optimum temperature for beetroot storage is 4–5 °C. The optimum relative humidity (RH) is between 80 and 90%, although very few seed producers have sufficient facilities to control the RH. A stacked tray or crate system is very suitable and can be coordinated with field operations at lifting and planting times. Where possible, the boxes of selected roots are air dried under cover before being stacked in the store.

Field storage: there are several versions of field storage, which include clamps and pits. The selected roots are arranged in pyramids or ridges on well-drained sites. In both these versions the roots are stacked 60–200 cm above ground level. The piles of roots are covered with straw, which is held in position by a covering of loose soil. Polythene sheets can also be used to exclude the rain and to give some frost protection, but care must be taken to avoid condensation. Straw funnels or chimneys in the ridges reduce the risk of condensation.

The roots are replanted in the spring, as early as local conditions allow, in rows 100 cm apart, with approximately 30 cm between roots within the rows. The roots must be set upright with their crowns at finished soil level. Some growers plant into a relatively loose soil, and pass a roller over the field after planting in order to firm the soil around the roots. This reduces drying out and is claimed to assist the early establishment of new fibrous roots.

In some areas, with a very well-organized and specialized vegetable seed industry, the roots are produced and stored by the seed companies and supplied to growers who plant them to produce the seed on a contract basis.

Seed to seed

In this system, a later sowing date is adopted compared with the root-to-seed system described earlier and the plants remain in the field all winter; therefore, this method is not suitable in areas with severe frosts. The actual sowing date depends on local climatic conditions, but August–September is the main time in the northern hemisphere. The sowing rate is approximately 12 kg/ha, which produces sufficient stecklings to plant 4 ha. The seed is sown in a four-row bed system with 25–30 cm between the rows, with a bed width of 110 cm. Sowing rates are adjusted according to seed type to give an optimum plant density in the beds of 200 plants/m².

The stecklings are transplanted in the early spring, but in some of the Mediterranean areas spring transplanting is difficult because the soil is very wet, and late autumn transplanting is adopted.

The optimum size of transplant is approximately 2.5–2.75 cm (weighing from 40 to 45 g). Although the trimming of transplants facilitates the operation, especially the long-rooted cultivars, the swollen tap root must not be cut.

The planting distances are 60 cm between the rows and 45–60 cm within the rows. In drier areas the transplants are frequently irrigated until established.

Flowering and pollination

The beetroot is a quantitative long-day biennial with a cold requirement for flower initiation. The detailed investigations into this requirement were first made by Chroboczek (1934). More recent work has demonstrated that exposure of the ripening seeds at low temperatures can reduce the subsequent low temperature requirement. This early vernalization sometimes contributes to the incidence of bolting in the first year.

The inflorescence emerges from the growing point relatively early in the spring. Some seed producers 'top' the flowering shoot when it is approximately 40–50 cm high. It is claimed that this increases seed yield by reducing the duration of flowering and concentrates the seed maturity period, which in turn reduces seed losses from shattering.

Beetroot flowers are predominantly wind-pollinated, although there is also some insect pollination by *Diptera* species.

Isolation

It is generally accepted that the pollen of *B. vulgaris* is wind-borne over relatively long distances, and sufficient isolation should therefore be ensured. Most authorities stipulate isolation distances of at least 500 m between cultivars of the same type (e.g. red globe) and at least 1000 m between different types of cultivar (e.g. between red globe and cylindrical types).

Beetroot is cross-compatible with the other subspecies of *B. vulgaris* (i.e. sugarbeet, mangolds, spinach beet and Swiss chard) and adequate isolation of the different seed crops has to be ensured. This is usually accomplished by a zoning scheme aimed at confining seed production of each of the different subspecies in separate geographical areas. Zoning schemes can either be voluntary agreements between seed producers, plant breeders and seed organizations, or they may be a legal requirement. The minimum isolation requirements between the different types of *B. vulgaris* vary from one authority or scheme to another, and may be 1 km, although the recommended distance in some states of the USA is at least three times this distance.

When seed of high genetic quality is required, or pollen contamination is suspected, the discard strip technique, described by Dark (1971), can be used. Although this technique was developed for sugarbeet seed production, the same principles apply with other wind-pollinated *B. vulgaris* types.

Roguing stages

The roguing of plants for beetroot seed production is considerably more thorough when the root-to-seed system is used. The seed-to-seed system does not allow the mature root characters to be observed.

Seed to seed
The main roguing is done at lifting and replanting, although plants which bolt prematurely can be removed before lifting:

1. Lifting: discard plants showing any incorrect leaf shape and/or colour, premature bolting, incorrect root shape or seed-borne pathogens.
2. Replanting: characters as described for lifting.

Root to seed
1. Before cutting tops for lifting: remove plants showing any incorrect leaf colour and/or morphology, early bolters and plants showing symptoms of seed-borne pathogens.
2. Lifted roots: discard roots which are not true to type. Shape, size, crown, base and surface corkiness should be taken into consideration.
3. Replanting: if roguing has been done in accordance with stage 2, no further roguing is required, although roots showing storage diseases should be discarded.
4. Bolting plants (before 'topping'): remove plants showing incorrect leaf shape, colour, vigour and seed-borne pathogens.

Improvement of basic stock

The production of basic stock seed is always done by the root-to-seed method. All the crop is rogued at all stages described for root-to-seed production, and in addition interior root characters are observed before replanting by cutting out a thin wedge of root flesh. This enables the degree of colour differentiation between the vascular and parenchyma tissues to be observed. Alternatively a cork borer is used to remove a sample of root flesh and the core examined. Unsatisfactory roots are discarded before planting the remainder. Some seed producers replace the wedge or core while others prefer to apply a fungicide dust to the exposed surfaces before planting.

In addition to this inspection of interior root characters, the discard strip technique can be used for the production of stock seed.

Harvesting

The harvesting of beet seed commences when the 'fruits' at the bases of inflorescence side shoots mature. By this stage the fruits have turned from green to brown. An additional check is to cut a few samples of ripe fruits transversely; unripe fruit are milky when cut and the ripe fruit are mealy. Seeds ripen successively from the bases of the side shoots to the terminal point. Care is needed to determine the optimum time for cutting because the immature seeds shrivel if cut too early, and if cut too late seeds are lost as a result of shattering.

Ripening beetroot stems tend to be prostrate rather than vertical. The method of cutting depends on the scale of operation. The large-scale producers in the USA use a swather, but for small-scale production, basic seed or larger-scale production where hand labour costs are relatively low the crop is cut with knives or hooks.

The cut stalks are left in windrows to dry and carefully turned once or twice. In areas where autumn rain is a problem, the cut stalks are tied in bundles and dried on 'four poles' placed in shocks (stooks) or alternatively transferred to the protection of open sheds. Large heaps of cut material are placed on tarpaulins or polythene sheets to avoid loss of seed from shattering. The cut material can take from 3 to 14 days to dry according to air temperature and rainfall.

Threshing

After drying, the materials are threshed by a stationary thresher or combine. The dry straw of beetroot seed is extremely brittle; it is therefore important to use a relatively low cylinder speed and air blast. Concave openings must be wide in order to avoid producing too many small pieces of straw as it is difficult to separate and remove these afterwards. There is relatively little chaff in beetroot material.

The final separation of beetroot 'seeds' from the small pieces of plant debris is done on a gravity separator.

Table 7.1. The main seed-borne pathogens of *Beta* species; these pathogens may also be transmitted to the crop by other vectors.

Pathogens	Common names
Alternaria alternate	Seedling rot, leaf spot
Cercospora beticola	Leaf spot (of warmer climates)
Colletotrichum dematium f. *spinaciae*	
Erysiphe betae	Powdery mildew
Fusarium spp.	
Perenospora farinosa f. sp. *betae*	Downy mildew
Pleospora betae	Blackleg, damping-off, leaf spot
Ramularia beticola	Leaf spot
Curtobacterium flaccumfaciens pv. *betae*	Silvering of red beet
Pseudomonas syringae pv. *aptata*	Bacterial blight, leaf spot, black streak, black spot
Arabis mosaic virus	
Tomato black ring virus	Beet ringspot virus
Raspberry ringspot virus	
Lychnis ringspot virus	
Ditylenchus dipsaci	Eelworm canker

Seed yield and 1000 grain weight

A satisfactory yield of beetroot seed in most areas of the world is approximately 1000 kg/ha, although up to twice this amount is usually achieved in the USA.

The 1000 grain weight of multigerm beetroot 'fruits' is approximately 17 g, rubbed and graded 'seed' has the lower 1000 grain weight of approximately 10 g, but as the size of the 'fruits' varies between seed lots these figures are only given as a guide.

Pathogens

The main seed-borne pathogens of *B. vulgaris* with common names of the diseases they cause are listed in Table 7.1.

Swiss Chard, Chard, Leaf Beet, Spinach Beet: *Beta vulgaris* L. ssp. *cycla* (syn. *Beta vulgaris* L. var. *vulgaris* L.)

This crop, which is another subspecies of *B. vulgaris*, has been developed and selected for its broad leaves and wide petioles. It is a biennial and is cross-compatible with the other types of *B. vulgaris* (beetroot, sugarbeet and mangolds).

The fresh market crop is grown commercially in Europe (especially France and Italy) and North America. It is also a popular garden vegetable in these areas, the Mediterranean, Asia and the higher elevations of the tropics and subtropics where it is grown in the dry season. The adoption of this crop for home gardens schemes is encouraged due to its nutrition value as a leafy vegetable.

Cultivar description of spinach beet

- Season of production: suitability for specific seasons, resistance to early bolting and low temperatures and adverse weather.
- Leaf blade: length, width, intensity of green colour, reflexing of margin, glossiness, blistering, anthocyanin absent or present.
- Petiole: length, width, colour, red, green, yellow green or bicolour.
- Relative width and length.
- Resistance to specific pathogens and pests.

(A detailed test guideline (106) is obtainable from UPOV; see Appendix 2.)

Agronomy

The soil and nutrient requirements and seed production methods are similar to beetroot. However, because the leaves are relatively large, a lower plant population is used per unit area. Seed is drilled at the rate of 6 kg/ha in rows 90 cm apart. Seedlings are singled and thinned to a final stand of 30–45 cm within the rows.

As with the production of beetroot seed, both seed-to-seed and transplanting systems are used. The transplanting system allows a better opportunity for plant inspection and is the only method used for the production of basic seed. Unlike beetroot, the transplants are not stored. Commercial seed is produced in areas of the USA where the winter climate is sufficiently mild for plants to survive in the field and be transplanted in the spring. In Europe, especially, southern France and Italy, the selected plants from a summer sowing are transplanted in the autumn and overwintered in unheated plastic structures. The plastic protection is removed in the spring. In cooler climatic areas the plastic is left on the structures.

Isolation

The minimum isolation distance between any similar leaf colour cultivar is 1 km. This is doubled for cultivars with different leaf colours. The recommendation or stipulation for Swiss chard isolation from other types of B. vulgaris is frequently at least 5 km in the USA, but depends on whether, or not, sugarbeet seed is produced in the area. The importance of sugarbeet seed production tends to dictate the isolation distances between the different types of B. vulgaris in the major seed producing areas of the world.

Roguing stages

Seed to seed

In this system roguing is done before the onset of winter and again in the spring. Early bolting plants are removed in the autumn and on both occasions the leaves and petioles are inspected for trueness to type. Rogue plants and those with seed-borne pathogens or pests are removed.

Root to seed (transplanted crops)

The plants are examined before lifting, while the typical leaf and petiole characters can be checked. Off-types and bolting plants are discarded before replanting in their winter situation. The plants are examined again at the end of the winter period for a final confirmation of morphological characters and freedom from seed-borne pests and pathogens. In Europe the selected plants are usually overwintered in protective structures (see Fig. 7.1).

Harvesting

The stages of ripeness for harvesting are similar to beetroot. Swiss chard inflorescences are from 2 to 3 m high, but after fertilization may become more horizontal and meshed together, therefore the seed crop is very prone to lodging. This is a similar situation to sugarbeet, and therefore a sugarbeet seed harvester with two cutter bars and a special windrower is very appropriate for large-scale

Fig. 7.1. Selected plants of Swiss chard that have overwintered under plastic.

harvesting operations. For smaller-scale production the crop is cut with knives or hooks (as described for beetroot) and dried in windrows.

The harvested material is threshed and seeds separated by the same process as described for beetroot.

Seed yield and 1000 grain weight

The seed yield is approximately 1.1 t/ha, with good yields of up to 2 t/ha reported by specialist seed producers in the USA.

The 1000 grain weight of Swiss chard is approximately 17 g.

Seed-borne pests and pathogens

The seed-borne pests and pathogens are the same as those listed for beetroot (see Table 7.1).

Spinach: *Spinacea oleracea* L.

The common name 'spinach' is given to different species of leafy vegetables in different parts of the world. Seed production of the so-called European spinach (*Spinacea oleracea*) is discussed in this section. 'African spinach' (*Amaranthus* spp.) is dealt with in Chapter 16. The species that is commonly known as Indian or Malabar spinach is *Basella alba* L. and is in the botanical family *Basellaceae*.

Spinacea oleracea was introduced into Europe from south-west Asia probably in the 14th century, and it later reached North America. It is now widely cultivated in all temperate regions. Cultivation in the tropics is not important except in the higher elevations.

The modern cultivars have been developed for their abundance of edible leaves. In the last four decades the crop has become important to processors, who market it as a canned purée and in frozen packs. It has also gained in popularity when produced as a small leaf salad crop.

Cultivar description of spinach

There are open-pollinated and hybrid cultivars and round or prickly seeded:

- Season: relative earliness, suitability for long-day sowing, resistance to early bolting.
- Leaf blade: approximate number of leaves before bolting, colour, shape, texture smooth or crinkled.
- Leaf petiole: length, pose, colour.
- Resistance to downy mildew and Cucumber mosaic virus.

(A detailed test guideline (055) is obtainable from UPOV; see Appendix 2.)

Soil pH and nutrition

The optimum soil pH is 6.0–6.8; if the pH is below 6.0, an appropriate dressing of a liming material is applied during soil preparations.

The general N:P:K recommendation is in the ratio of 1:2:2 applied during the final stages of seedbed preparation. Supplementary top dressings of nitrogen are given before and after bolting. Spinach plants are prone to lodging after flowering commences, and applications of nitrogen should therefore only be applied according to local experience and the amount of nitrogen lost by leaching.

Soils known to be infected with fusarium or verticillium wilt pathogens must not be used for seed production.

Sowing

In the USA, sowing is done either in the autumn or spring. Canadian producers sow only in the autumn. In Europe, sowing is done at either time according to location and hardiness of the cultivars.

Seed is sown at the rate of 6 kg/ha, in rows 45–60 cm apart. Seedlings are not normally thinned except for the production of basic seed.

Flowering and flower type

Spinach is a typical long-day plant. Flowering occurs in unvernalized plants, but is hastened by previous chilling.

Populations of spinach are composed of plants that are male, female or hermaphrodite. Male plants tend to flower before female plants. There also tends to be a correlation between the dioecious plants, leaf size and number; male plants produce fewer and smaller leaves before flowering whereas female plants produce more and larger leaves before flowering.

Pollination and isolation

Spinach is mainly wind-pollinated. Recommended isolation distances are therefore up to 1000 m in some countries, although some authorities stipulate only 500 m for production of commercial seed of cultivars within the same type (e.g. leaf type and seed type).

Roguing stages

1. Before main flowering (when the plant has formed a rosette), remove non-rosetting and early flowering male plants that are not true to type.
2. When flowering has commenced, as for stage 1.

Plants that are infected with specific seed-borne pathogens, e.g. mosaic (cucumis virus 1), are removed during roguing.

Hybrid seed production

The ratio of female-to-male rows depends on the potential pollen production of the male parent, and is usually 6:2 or 14:2.

Specific instructions for roguing and harvesting should be provided by the maintenance breeder. Roguing for hybrid seed production includes the removal of male plants from the female rows, this is usually done twice to ensure complete removal of the males from the female rows. Some hybrid lines are produced by crossing round-seeded and prickly seeded parents, and this enables the seed products of the parents to be harvested together and subsequently separated during processing (Sneep, 1958).

Harvesting

The ripening crop seed is subject to loss from shattering and birds. In the large-scale seed-producing areas with relatively calm and dry weather conditions, the crop is harvested with a combine when the plants are dry and the majority of the seeds are mature. Otherwise the crop is cut and placed in windrows to dry as soon as the plants start to dry out and the earliest seeds are mature. An approximate guide to this stage is when the later ripening plants start to become yellow. Material left to dry must be stacked on sheets to avoid loss from shattering.

Threshing

Spinach seed from cut and dried plants is threshed with a small-drum thresher or a cereal thresher. In the latter case the recommended drum speed is about 700 rpm, and the concaves are set relatively wide to minimize the amount of broken stalks. In addition to separating the seeds from the plant, the threshing operation breaks up the clusters of seeds.

Seed yield and 1000 grain weight

The generally accepted seed yield is approximately 800 kg/ha, although yields of up to 2000 kg are reported. The yield of hybrid cultivars per unit area is very similar to open-pollinated cultivars even allowing for discarding the male lines.

The 1000 grain weight of spinach is approximately 10 g.

Pathogens

The main seed-borne pathogens of spinach are listed in Table 7.2.

Table 7.2. The main seed-borne pathogens of spinach species; these pathogens may also be transmitted to the crop by other vectors.

Pathogens	Common names
Cladosporium variabile	Leaf spot
Colletotrichum dematium f. *spinaciae*	Anthracnose
Colletotrichum spinaciicola	
Fusarium oxysporum f. sp. *spinaciae*	
Perenospora farinosa f. sp. *spinaciae*	Downy mildew, blue mould
Phyllosticta spinaciae	Leaf spot
Rhizoctonia solani	Damping off
Verticillium dahliae	
Verticillium sp.	Wilt
Spinach Latent Virus	

Further Reading

Spinach

Shinohara, S. (1984) *Vegetable Seed Production Technology of Japan, Elucidated with Respective Variety Development Histories, Particulars*, Volume 1. Tokyo, Japan, pp. 123–142.

8   *Asteraceae* (formerly *Compositae*)

Although this family contains a large number of genera, there are relatively few that are of major importance as cultivated vegetables. The main vegetables that are produced from seed are:

Lactuca sativa L.	Lettuce
Cichorium endiva L.	Endive
Cichorium intybus L.	Chicory, witloof chicory

There are many species of local importance which are used as leafy vegetables and include: *Tragopogon porrifolius* L., salsify or vegetable oyster plant, which is cultivated as a minor root vegetable in some temperate areas including northern Europe; *Taraxicum officinale* Wigg., dandelion (which is also a common weed in parts of the world, including Europe); and *Lactuca indica* L., cultivated as a leafy vegetable in Asia, especially China, Indonesia and The Philippines.

Lettuce: *Lactuca sativa* L.

Origin and types

The cultivated lettuce *Lactuca sativa* is generally thought to be derived from the wild species of *Lactuca serriola* L. There is a very wide range or variation within the cultivated forms of *L. sativa*, which are divided into four types. The divisions, based on morphological characters, are:

1. *Lactuca sativa* var. *capitata* L.: the cabbage or head lettuce, which is generally subdivided into crispheads (iceberg) and butterheads. The crispheads have firm hearts produced by the close overlapping of course-veined brittle leaves with prominent midribs.

The butterheads have relatively soft textured leaves with a greasy appearance. As in the crispheads, the leaves of mature plants form a heart, although less firm than the crispheads.

2. *Lactuca sativa* var. *longifolia* Lam.: the cos or Romaine lettuce. The upright, relatively narrow, crisp textured leaves form a closed head.

3. *Lactuca sativa* var. *crispa* L.: the 'leaf' or curled lettuce does not form a heart but has a loose head of leaves. Some of the cultivars have very curly and fringed leaves.

4. *Lactuca sativa* var. *asparagina* Bailey (syn. var. *angustana* Irish): all the forms of this group have typical fleshy stems which are the main culinary attraction, especially in Asia. The group may also be referred to as 'stem lettuce', a typical cultivar is Celtuce. Some of the members of this group have a light-grey leaf colour.

There is an increasing interest and demand in some markets for lettuce marketed as' baby leaves', especially for the pre-packed salad trade.

Systems for the classification and identification of lettuce cultivars have been documented by Bowring (1969) and Rodenburg and Huyskes (1964). Classifications according to seedling characters only have been based on work by Rodenburg (1958), and while extremely useful for 'Distinctness, Uniformity and Stability' tests, the characters are useful for selection or roguing only if observations are made before about the fifth leaf stage. Some of the seedling characters such as absence or presence of anthocyanin cannot be observed after the first true leaf has emerged. Clarkson and George (1985) have examined the application of seed and seedling characters up to the third leaf for the identification of lettuce cultivars, examples of the third leaf stages of six cultivars are shown in Fig. 8.1.

Tom Thumb   Paris White   Continuity   Lollo Rossa   Oak Leaf   Salad Bowl

Fig. 8.1. Third leaf of six lettuce cultivars showing range of morphological types.

Cultivar description of lettuce

- Seed: colour white, yellow or black.
- Morphological head type: butterhead, cos or crisp; hearting or leafy.
- Season of use: summer, outdoor or protected cropping, winter hardiness.
- Seedling or young plant characters at 3–4 leaf stage (see Fig. 8.1).
- Anthocyanin pigmentation: leaf colour and form.
- Mature plant: leaf lustre, degree of blistering.
- Time to maturity, relative duration mature plant holds for market before bolting.
- Resistance to specific races of *Bremia lactucae* (downy mildew).
- Resistance to Lettuce Mosaic Virus (LMV).

(A detailed Test Guideline 013 is obtainable from UPOV, see Appendix 2.)

Soil pH and nutrition

This crop should be grown in a soil which has a pH of at least 6.0. Because lettuce is susceptible to calcium deficiency, it is preferable to adjust the pH to 6.5 by adding a suitable liming material during site preparation.

The lettuce seed crop responds to the same pattern of nutrients as the market crop and seed producers generally use a base dressing of 3:2:2 of N:P:K. Nitrogen plays an important role in total seed yield which responds to up to 80 kg nitrogen/ha. However, excess nitrogen tends to produce a loose atypical head which is difficult to confirm as true to type when roguing or selecting. Nitrogen can be applied during seedbed preparation but supplementary top dressings or foliar sprays (using urea) should be applied during the crop's growth. This is especially useful if there is likely to have been significant amounts of leaching. Extra applications of nitrogen when the plants have commenced bolting will increase seed yield.

Sowing and young plant establishment

The seed is drilled into rows 50–60 cm apart at a rate of 1.5–2 kg/ha. The young plants are thinned to between 25 and 30 cm apart within the rows. Plants raised in modules are planted 25 cm apart with 50–60 cm between the rows.

Irrigation

The lettuce seed crop responds directly to irrigation. In addition to increasing total plant weight in the vegetative stage, a satisfactory supply of water assists in the increase of total seed yield, although there is a tendency for seed crops receiving frequent irrigation to be up to 5 days later in maturity. The significant gain in seed yield is generally considered to outweigh the possible disadvantage of a slight delay to optimum harvest time of the seed.

The influence of water stress at different lettuce growth stages was investigated by Izzeldin *et al.* (1980). They reported that the highest seed yield was obtained when the mother plants were subjected to a moderate water stress during the vegetative stage but followed by adequate available water during anthesis. Water stress during anthesis resulted in a reduction in seed yield.

In areas where the total water requirement is applied through overhead irrigation systems, growers generally maintain a drier regime when seed ripening has commenced. Overhead irrigation is detrimental once seed has begun ripening because the impact of the water droplets causes ripe seed to fall. In some areas, or conditions, irrigating late in the life of the crop encourages weed growth which can make harvesting difficult and increase contamination of the harvested seed with weed seeds.

Flowering and pollination

The physiology of lettuce has been reviewed by Wien (1997b), who reported that lettuce is a quantitative long-day plant and that it can also be seed vernalized. The vernalization requirements of some lettuce cultivars and other *Lactuca* species have been investigated by Prince (1980). As there are types which heart in response to day length, it is important to grow individual cultivars under the conditions prescribed by the breeder in order to confirm the morphological characters. Waycott (1993) has discussed the effect of photoperiod on the transition to flowering in lettuce genotypes.

The inflorescence of lettuce, which is called a capitulum, contains approximately 24 florets. These are highly developed in favour of self-pollination and the crop is therefore largely self-fertilized. However, some cross-pollination can take place between lettuce cultivars and also between cultivated lettuce and some wild *Lactuca* species. For example, the cultivated lettuce *Lactuca sativa* is cross-compatible with *Lactuca serriola*, which is a wild plant and fence-line weed, especially in countries around the Mediterranean and parts of Asia. If mechanical contamination of the lettuce seed crop with seeds of wild *Lactuca* species occurs, it presents insurmountable problems during the subsequent seed processing.

Isolation

For the production of certified seed a period of 3 years should have elapsed from a previous lettuce seed crop produced on the same site or 2 years from a previous market crop. Some authorities make an exception to this minimum isolation period when the soil or substrate has been effectively fumigated or partially sterilized. This exception is especially useful when lettuce seed is being produced in protected structures or glasshouses.

Although up to 5% cross-pollination has been observed in lettuce in some areas, most authorities regard it as a self-pollinating crop and only specify a physical barrier (e.g. adjacent sections of greenhouses) of a minimum of 2 m between different cultivars.

Roguing and selection

There are three main stages for roguing and selection, which are:

1. The young plant during the four- to six-leaf stage.
2. The mature plant at time of heading.
3. After bolting has commenced.

There is an additional earlier stage when the plant is still a seedling, up to the third leaf. In practice this can only be used prior to planting, when the plants have been raised in modules or if relatively few seedlings are being examined (see Fig. 8.1).

The most important roguing stage is at the time of heading; in practice this is usually the only stage at which commercial seed crops are rogued.

The features used in the assessment of trueness to type of lettuce seed crops are mainly based on the morphological characters observed during the vegetative stages of plant growth up to and including hearting (or heading in the case of non-hearting cultivars) and are listed above under cultivar description.

Seed stalk emergence

The seed stalk does not always emerge readily from the lettuce heart. Although this is not a major problem with the butterhead types, it can be with the crisp-heads because many of the cultivars in this group form very firm heads which are mechanical barriers to the emerging inflorescence. The poor emergence results in the development and elongation of basal shoots, relatively few flowering shoots on the main axis which becomes distorted and a reduction of the potential seed yield. The general tardiness of bolting of firm-headed lettuce also results in an increase in the mature plant's exposure to pathogens such as *Botrytis cinerea.*

Improving seed stalk emergence

Traditionally, several mechanical methods have been used to assist the emergence and elongation of the flower shoots. These have included deheading, slashing and quartering of the heads with special tools to assist the breakage of the solid heart without damaging the unexposed growing point; some lettuce seed producers refer to this operation as 'racing'. Another simple method is to give a sharp blow with the palm of the hand to the top of the lettuce head. This fractures the leaf bases and, following removal of the detached head, the growing point remains unharmed. An alternative is to manually peel off the leaves around the heart, although this is only really practical when relatively few plants are to be used for seed production. The timing of the mechanical or manual operations is important; if left too late, the extended flowering shoot inside the heart will be damaged.

Growth-regulating chemicals have been used to promote bolting in lettuce. Gibberellic acid has been applied before hearting to promote early bolting, thus avoiding the problems associated with flower shoot emergence. The application of aqueous solutions of gibberellic acid at concentrations between 20 and 500 ppm have been used for the butterhead types (Harrington, 1960). However, the application of chemicals to lettuce before the hearts have been completely formed in order to assist flower-stem emergence does not enable the seed producer to confirm the plant's morphological characters at normal market maturity and can therefore adversely affect the selection or roguing efficiency.

Wurr *et al.* (1986) included the use of gibberelin 4 plus 7 in their studies of the effects of seed production techniques on seed characteristics, subsequent seedling growth and crop performance of the crisp lettuce cultivar 'Pennlake'. They concluded that the major differences between commercial seed lots of lettuce result from differences in the environment rather than the seed production technique.

Seed development and harvesting

The succession of inflorescences during anthesis subsequently provides a steady sequence of ripe seed. The length of time from flowering to ripe seed produced on an individual capitulum is 12–21 days, depending on environment. High temperatures increase the rate of development and ripening.

The influence of post-flowering temperature on seed development and subsequent performance in crisp lettuce, cv. Saladin, was investigated by Gray *et al.* (1988b). They reported that seed yields per plant increased from a mean of 15–27 g, and fell to 20 g with progressive temperature increases (in 16 h day and 8 h night) of 20/10°C, 25/15°C and 30/20°C; they also included evaluations of resulting seed quality. In general, seed produced at 25/15°C exhibited a greater variation in numbers of seeds per floret, cell numerical volume density, seed weight, times of seedling emergence and seedling and mature head weight than seed produced at the lower temperatures.

If the seed producer waits for the development and ripening of the seeds from the later inflorescences, the earlier ripened ones will probably have been shed and lost. It is, therefore, general practice to cut and harvest the seed when an estimated 50% of seed heads are ready on a typical sample. The stage of ripeness at which the pappus is fully developed and dry is referred to as 'feathering', as illustrated in Fig. 8.2.

The standing lettuce seed crop can be hand- or machine-cut and left in windrows. A machine which minimizes shattering should be used for the cutting process. There is less shattering if the plants are cut with the dew on them. The cut material is normally left in windrows for up to 5 days, but in very arid areas the seed can be extracted the same day as cutting. After windrowing the

Fig. 8.2. Lettuce with approximately 50% of seed ready for harvest.

seed is extracted in stationary threshers or passed through a combine. If the seed is combined direct from the standing crop, many small pieces of wet plant debris, such as fragments of leaf bract, will be mixed with it, causing an undesirable increase in seed moisture unless the material is quickly dried.

In production areas, with plenty of available labour, the seed can be harvested from single plants by shaking their heads into a canvas bag or sack. If this is done every 2 or 3 days the maximum amounts of seed are collected.

Seed cleaning

Pre-cleaning vegetable seeds of *Asteraceae* is an important operation because the weight of chaff in mechanically harvested seed can be as high as one-and-a-half times the weight of the seed itself. The initial cleaning can be done with an air-screen machine. There is usually some plant material such as small pieces of flower stem remaining in the sample which can be removed by passing the seed through a disc separator or an indent cylinder.

Seed yield and 1000 grain weight

A satisfactory seed yield for hearting lettuce under good conditions is between 0.5 and 1 t/ha.

The 1000 grain weight of lettuce is from 0.6 to 1.0 g according to the cultivar.

Indexing for lettuce mosaic virus (LMV)

The macroscopic symptoms of LMV are observed as a clearing between the veins when a portion of leaf from an infected plant is held up to the light. Plants which become infected relatively early in their life are usually stunted and frequently fail to heart. When seeds are saved from infected plants, up to 15% of the seeds carry the virus in their embryos. Plants which are infected as a result of seed transmission act as reservoirs from which sap-sucking insect vectors transmit the virus to other plants in the same or neighbouring crops. The aphid *Myzus persicae* is especially important as a vector of this virus.

In addition to ensuring satisfactory isolation from other lettuce crops it is important to rogue out infected plants as soon as the symptoms are observed. Basic seed stocks can be produced in insect-free structures or, in some parts of the world such as Australia and California, at temperatures too high for aphid attack. There are other viruses of lettuce that cause similar mosaic-like symptoms but they have not been found to be seed-borne.

Mosaic-indexed seed is produced under strictly controlled conditions which incorporate satisfactory isolation, roguing and virus vector control. The resulting seed is then subject to 'mosaic indexing'. In the standards for seed health of selected vegetable crops for certification in the UK, this involves growing on a sample of 5000 seeds derived from the seed crop in a cool insect-proof greenhouse or structure and noting the incidence of LMV. A basic or certified lettuce seed stock which has been found to produce 'nil infection' from this screening is said to be 'mosaic-indexed'. This is an example of how the seed producer can assist the grower to avoid problems arising from seed transmitted pathogens.

Pathogens

The main seed-borne pathogens of *Lactuca sativa* with the common names of the diseases they cause are listed in Table 8.1.

Witloof Chicory: *Cichorium intybus* L.

This is a crop grown as a winter vegetable by specialist producers especially in northern Europe. The market crop is produced by forcing the 1-year-old dormant roots.

The cultivars are normally classified according to their relative earliness when forced, as there are very few other clear distinctions. The time to maturity of a specific cultivar depends on its vernalization requirement. Rodenburg and Huyskes (1964) developed a scheme for classifying witloof chicory according to the length of the core, i.e. the stem in the head. They found that the longer the core, the earlier is the cultivar. This is a correlation with the vernalization requirement, as the successive cultivars have an increasing cold requirement. The cultivars with the lower cold requirement will tend to bolt when sown relatively early in their first year.

Table 8.1. The main seed-borne pathogens of *Lactuca sativa*; these pathogens may also be transmitted to the crop by other vectors.

Pathogens	Common names
Alternaria dauci	
Marssonina panattoniana	Ring spot, anthracnose
Pleospora herbarum f. *lactucum*	Leaf spot
Sclerotinia sclerotiorum	Drop, watery soft rot
Sclerotium rolfsii	Southern blight
Septoria lactucae	Leaf spot
Pseudomonas cichorii	Leaf blight
Arabis mosaic virus	
Lettuce mosaic virus	
Lettuce yellow mosaic	
Tobacco ringspot virus	
Tomato black ring virus (ringspot strain)	

There are also types which have been developed as a salad crop (referred to as 'radicchio', or salad types) and are available as cultivars which form a heart or semi-heart with a range of leaf colours including red, pink and variegated.

C. intybus is also used for the production of roots which are dried, ground and used as an additive to coffee. There are no records of cultivars specially developed for this use. Baes and Van Cutsem (1992) have reported a method of identifying seed lots of closely related chicory 'varieties' by native polyacrylamide gel electrophoresis and subsequent leucine aminopeptidase and esterase staining of bulked seed sample extracts.

Agronomy and seed production

The seed crop is grown as a biennial, and subsequently treated similarly to lettuce. Some seed companies have developed F1 hybrid cultivars of witloof chicory.

This crop is largely cross-pollinated, and there should be a minimum of 1000 m between seed crops of different cultivars.

Seed yield and 1000 grain weight

The average chicory seed yield is approximately 1 t/ha.

The 1000 grain weight of chicory is approximately 1.4 g.

Table 8.2. The main seed-borne pathogens of
Cichorium intybus; these pathogens may also be
transmitted to the crop by other vectors.

Pathogens	Common names
Alternaria cichorii	Black leaf spot
Botrytis cinerea	Grey mould
Gibberella avenacea	
Rhizoctonia solani	
Chicory yellow mottle virus	

Pathogens

The main seed-borne pathogens of *C. intybus* are listed in Table 8.2.

Endive: *Cichorium endiva* L.

This is a popular salad crop in northern Europe, especially France and Italy,
where seed companies offer a range of cultivars. Some of the traditional market gardens grow the plants to maturity and then 'blanch' them by excluding
the light. However, many of the present day cultivars have been developed for
production and marketing in a similar way to lettuce, i.e. without 'blanching'.

Origins and types

It is not clear if the cultivated derivatives of this species are of Indian subcontin-
ental or Mediterranean origin but selections of the species have been cultivated
in both areas for centuries (Hedrick, 1972). Some forms of *C. endiva* are
annuals, but seeds of the main cultivars offered by continental seed companies
are produced as a biennial seed crop.

Cultivar description of endive

- Season: relative time of maturity, summer or autumn type, suitability for
  protected cropping.
- Size: relative size of mature head.
- Growth form erect, semi-erect or prostrate.
- Leaf: colour green, degree of white to mature head.
- Shape broad-leaved or curled.
- Flower: colour white, pink, blue or violet blue.
- Time of maturity: early, medium or late.
- Standing ability at market maturity; bolting: early, medium or late.

(A detailed test guideline (118) is obtainable from UPOV; see Appendix 2.)

Agronomy

The seed crop has the same soil requirements as lettuce. Seeds are sown in the late summer or autumn in frost-free areas. As the plants are more vigorous than lettuce when flowering, lower plant densities are used per unit area, with rows 75 cm apart and plants thinned to 25–50 cm apart according to the cultivar's vigour.

Isolation

This species is considered to be mainly self-pollinating, and the isolation requirements are therefore similar to lettuce, but different types should be isolated by a minimum of 50 m.

Roguing stages

The plants are examined as described for lettuce during the rosette and mature-head stages, taking into account the range of characters included in endive cultivar descriptions.
    The two main roguing stages are:

1. Young plants in the rosette stage. Remove plants which are bolting early. Check the general morphology and leaf characters.
2. Market maturity stage. Check the general morphology and leaf characters. Remove plants bolting early. Examine for tolerance to adverse weather.

Seed yield and 1000 grain weight

The average seed yield is approximately 500 kg/ha, although yields in Italy reach 1 t/ha.
    The 1000 grain weight of endive seed is approximately 1.3 g.

Seed-borne pathogen

The main seed-borne pathogen of *C. endiva* (endive) is *Alternaria cichorii*, commonly referred to as black leaf spot.

Further Reading

Lettuce

Shinohara, S. (1984) *Vegetable Seed Production Technology of Japan, Elucidated with Respective Variety Development Histories, Particulars*, Volume 1. Agricultural Consulting Engineer Office, Tokyo, Japan, pp. 143–174.

Lettuce, endive and chicory

Ryder, E.J. (1999) *Lettuce, Endive and Chicory*. CAB International, Wallingford, UK.

9 *Cruciferae*

This family contains several important vegetables and includes:

Brassica oleracea L.	The brassicas or cole crops
var. *acephala* DC	Kale
var. *capitata* L.	Cabbage
var. *botrytis* L.	Cauliflower
var. *italica* Plenck.	Sprouting broccoli
var. *gemmifera* Zenker.	Brussels sprouts
var. *gongylodes* L.	Kohlrabi
Brassica campestris L.	Turnip and related crops
ssp. *rapifera* Metz.	True turnip
ssp. *chinensis* Jusl.	Chinese mustard, pak-choi
ssp. *pekinensis* (Lour.) Rupr.	Chinese cabbage, pe-tsai
Brassica napus L. var. *napobrassica* Rchb.	Swede
Brassica juncea (L.) Czern. and Coss.	Indian or black mustard
Brassica nigra (L.) Koch	Black mustard
Sinapis alba L.	White mustard
Lepidium sativum L.	Garden cress
Raphanus sativus L.	Radish
Rorippa nasturtium aquaticum (L.) Hayek	Watercress

In addition to the above vegetable crops, there are some economically important agricultural crops including oilseed and fodder species. Some of the *Brassica* oilseed crops have become important for biofuel production.

Cole Crops

This group of *B. oleracea* types, often referred to as cole crops, is generally considered to include Brussels sprouts, cauliflowers (including broccoli), kales, kohlrabi and cabbage (but not *B. campestris* subspecies, the Chinese cabbage types). The general history of the main types has been discussed by Thompson (1976). In addition to including important vegetable crop cultivars, there are also cabbage, kale and kohlrabi cultivars that are important agricultural crops for fodder and stock feed.

Flowering

Flower initiation in most cole crops is dependent on a low temperature stimulus. This vernalization requirement varies between the different types and between cultivars within the individual types. The physiology of the main vegetable cole crops, including flowering, has been reviewed by Wien and Wurr (1997).

Cultivar description of cabbage

- There are open-pollinated and F1 hybrid cultivars.
- Season of market crop and use; fresh market, processing or storage.
- Morphological type: head shape round, flat or pointed.
- Outer foliage: colour, pigmentation, waxiness of cuticle, texture.
- Resistance to specific pathogens, including race 1 of *Fusarium oxysporum* f. sp. *conglutinans*.
- Resistance to splitting and early bolting.

(A detailed test guideline (048) is obtainable from UPOV; see Appendix 2.)

Cauliflower: *Brassica oleracea* L. var. *botrytis* L.

Cultivar description of cauliflower

- There are open-pollinated and F1 hybrid cultivars.
- Season of production: suitability for overwintering, frost resistance, low-temperature requirement for curd initiation.
- Leaf: attitude, shape, midrib and vein characters. Curd protection by leaves.
- Curd: configuration, colour, relative size.
- Flower: colour; yellow or white.
- Resistance to race 1 of *Fusarium oxysporum* f. sp. *conglutinans*.

(A detailed test guideline (045) is obtainable from UPOV; see Appendix 2.)

Brussels Sprout: *Brassica oleracea* L. var. *gemmifera* Zenker

Cultivar description of Brussels sprout

- There are open-pollinated and F1 hybrid cultivars.
- Season of production; suitability for specific market outlet (e.g. freezing or fresh market); single harvest or continuous picking.

- Suitability for mechanical harvest.
- Vegetative characters: height, leaf shape, rugose, colour, petiole length.
- Sprouts ('buttons'): relative size, colour and firmness.
- Resistance to *Fusarium oxysporum*.

(A detailed test guideline (054) is obtainable from UPOV; see Appendix 2.)

Kohlrabi: *Brassica oleracea* L. var. *gongylodes* L.

Cultivar description of kohlrabi

It is important to note that the UPOV description refers to the marketable part of the plant as the 'kohlrabi', while many seedsmen and growers refer to it as the 'bulb', although botanically it is a swollen stem.

- There are open-pollinated and F1 hybrid cultivars.
- Season and uses; suitability for protected cropping; resistance to early bolting.
- Bulb characters: colour; red, white, purple; extent of pigmentation if speckled.
- Shape: globe or flat.
- Internal quality: degree of fibre (if tendency to be fibrous).
- Foliage: short or tall, pose of leaves, relative length of petioles.

(A detailed test guideline (065) is obtainable from UPOV; see Appendix 2.)

Seed Production of the Cole Crops

Soil pH and nutrition

The optimum pH for cole crops is 6.0–6.5 and any required liming material should be applied during soil preparation. Acid soil conditions not only affect the availability of micronutrients such as molybdenum, but also increase the incidence of club root (*Plasmodiophora brassicae* Woron.).

The N:P:K ratio of nutrients applied during preparation varies widely between different seed production areas, but the general recommendation is 1:2:2. High nitrogen levels result in 'soft' plants that are less able to withstand frost, and for this reason top dressings of nitrogen are only applied in the spring. The spring application also replaces nitrogen leached during the winter. Where deficiencies of available manganese, boron or molybdenum exist, suitable additions of the appropriate materials should be applied. Cauliflowers are especially susceptible to molybdenum deficiency that causes the condition known as 'whiptail'. A solution of sodium molybdate is applied to the young plants prior to planting out in areas where this is likely to occur. Boron deficiency causes 'hollow-stem' or corky patches on the stems of cole crops and in cauliflowers can also be responsible for the browning of parts of the curds.

Irrigation

Most references in the literature relating to the irrigation of cole crops refer to their vegetative stage. Increasing the available water when the plants are in the vegetative stage will increase the amount of foliage produced but no work has been reported on the effects of water availability on seed yield. However, cauliflower curd size is significantly increased by supplementary irrigation 3 weeks before the market crop is normally harvested. Water stress in the cole crops increases the thickness of the waxy cuticle and produces plants with a relatively blue-green pigmentation. These changes frequently occur under tropical or arid conditions and may increase the difficulty of confirming some foliage characters during roguing or selection.

Plant production, sowing and spacing

Cabbage, cauliflower and Brussels sprouts plants are normally raised in seedbeds and planted out. The sowing dates depend on the local custom and experience with specific crops. Sowing is timed so that the cultivar receives sufficient vernalization (if required), and to allow for the plants to survive hard weather and reach anthesis and subsequently produce seed in suitable weather conditions. Approximately 60 g of seed thinly sown in drills 20 cm apart produce ten seedlings per 30 cm, which in turn will produce approximately 10,000 plants, subject to satisfactory germination and seedling emergence. Some of the specialized cauliflower seed producers raise transplants by sowing in a greenhouse and pricking off the seedlings into 7 cm pots or similar containers.

Young plants from the seedbeds or containers are planted out when they have reached the 5–7 leaf stage. Obvious off-types, blind plants and any showing signs of serious pathogens are discarded at this stage.

The final spacing depends on the vigour of the cultivar, but a useful criterion is to ensure that the plant density is as high as possible but there should be sufficient space to allow inspectors to examine the crop. Rows are usually 60 cm apart, although some of the smaller cabbages are planted in rows 40 cm apart. The distance between the plants within the rows is usually 30–60 cm. In some seed production areas a 'bed' or 'set' system is of up to four rows is used with a wider space between the beds.

Some seed producers earth-up cabbage and Brussels sprouts plants after they are established in their final stations. This is done as part of the cultivation programme to control weeds but it also reduces the risk of plants toppling over when they become heavier.

Kohlrabi is usually direct drilled for seed production as a seed-to-seed system, although transplanting is used for stock seed production. Unlike the other cole crops discussed above, it is possible to vernalize the seed of some quick bolting cultivars of kohlrabi. The treatment is given before a spring sowing: the seeds are pre-soaked in water for approximately 8 h, and then spread on moist sheets of filter paper at 20–22°C. Between 70 and 90% germination occurs in 24 h at these temperatures. The moist seeds are stored at −1°C for 35–50 days (Nieuwhof, 1969). After thawing slowly the seeds are sown direct into the field. The use of this technique does not allow for roguing against early bolters and it should not be used for production of stock seed.

Techniques for individual subspecies

Cabbage

If seed is to be produced from hearted cabbage it is usually necessary to incise the mature heads after checking trueness to type. There are several ways of achieving this but all have the objective of allowing the flower stalk to emerge unhindered by the mechanical barrier of the tightly folded leaves that encase it. A quick method is to cut the top of the head in the form of a cross, but the growing point must not be damaged. Figure 9.1a illustrates cabbage heads immediately after cutting and Fig. 9.1b shows the emerging flower stems after approximately 2 weeks.

Cauliflower

The plants are left *in situ* when the curds are expected to form at a time of year when subsequent fine weather for seed production is predicted. This is the timing with early summer cauliflowers. Techniques for dealing with selected plants at other times of the year are discussed under basic seed production techniques. The seed production of tropical and subtropical cauliflowers has been discussed by Lal (1993b); production and tests to guarantee seed quality in heat stress in Italy have been described by Barbuzzi and Silviero (1998).

Brussels sprouts

The terminal growing points are removed from plants after the final roguing. This encourages the development of flowering shoots from the lateral buds (i.e. 'sprouts') thereby increasing the total seed yield and uniformity of seed maturity. Some seed producers remove some of the lower sprouts.

Kohlrabi

Overwintered crops are either left in the ground, protected from frost by earthing over or covering with a suitable material; otherwise they are lifted and stored in frost-free conditions. The stored roots are replanted in early spring.

Pollination

Most cole crops are predominantly cross-pollinated, although some self-pollination can occur. The level of self-pollination is greatest in the summer cauliflowers (Watts, 1980). Bees and *Diptera* species are the main pollinating agents of *Brassica* species. Weiring (1964) has reviewed the use of insects for the pollination of *Brassica* crops. Natural pollinating insect populations can be supplemented by placing bee hives adjacent to *Brassica*

(a)

(b)

Fig. 9.1. (a) Cut cabbage heads. (b) Flowering shoots emerging.

seed crops for the duration of anthesis. Work by Faulkner (1962) showed blowflies to be very efficient pollinating agents when plants are flowering in confined places.

The presence of pollen beetle can be a significant problem in *Brassica* seed crops, but care must be taken in the use of chemicals for their control in order to avoid harming pollinating insects.

Isolation and previous cropping

Most authorities consider it important to have a greater recommended distance (up to 1500 m) between different types of *B. oleracea*, e.g. cabbages and kohlrabi, than between different cultivars of the same type, e.g. two cabbage cultivars (up to 1000 m). There are zoning schemes for the *Brassica* spp. in some countries.

Some of the common weeds of *Cruciferae* are alternative hosts to important seed-borne pathogens such as *Alternaria brassicola* and *Phoma lingam*, therefore care must be taken to avoid sites where these may present problems. Other problems of weed seed contamination in the harvested seed lot can be created by the presence of some weed species if they are allowed to go to seed within the scheduled seed crop, these include *Galium aparine* (cleavers) and *Chenopodium album* (fat-hen).

Roguing stages

Cabbage

1. Planting out or before heading: check for general foliage characters.
2. When heads have formed on most plants in the crop: check head characters, including shape, relative size and firmness. Discard plants that are either too early or too late for the cultivar.

Cauliflower

1. Before normal curding period: reject precocious or button curds that develop before the normal maturity period of the cultivar. Foliage; check that pose and other characters of leaves are true to type.
2. Curd at market maturity: check that colour, absence of bracts, absence of riciness, solidity, shape and form of curd and leaf protection are according to the cultivar.

Brussels sprouts

1. Planting out, or before significant button formation commences: check for general plant vigour, relative height and foliage characters including colour.
2. When the lower 'buttons' or buds reach normal market maturity: check general plant characters, productivity and quality of buttons. Discard plants that are not within the maturity of the cultivar.

Kohlrabi

1. During the final thinning: remove any plants that are bolting early, and also plants that are not true to type for foliage characters, vigour or pigmentation.
2. At normal market maturity of kohlrabi ('bulb'): remove early bolting plants, check that the kohlrabi ('bulb') and foliage characters and also pigmentation are true to type.

Hybrid seed production

The development of hybrid cultivars of Brussels sprouts has had a very significant impact on its commercial production; hybrid cultivars have also become important in cabbage, cauliflower, kale and kohlrabi. Sporophytic self-incompatibility has been utilized for the production of F1 hybrid cultivars in all types of *B. oleracea* (Riggs, 1987). The ratio of female-to-male parent is usually 1:1 or 2:1 (Takahashi, 1987). Recommendations of maintenance breeders must be strictly adhered to when producing seed of hybrid cultivars.

The incidence of sibs in the production of F1 hybrid seed sometimes presents problems; Fitzgerald *et al.* (1997) have described a technique for determining sib proportion and aberrant characterization in hybrid seed using image analysis.

Harvesting

All of the cole crops have a strong tendency to shattering of their seed pods. It is therefore important that appropriate action is taken to secure the potential seed crop before any is lost. Work by Still and Bradford (1998) has indicated that population-based hydrotime and ABA-time models could be used to assess physiological maturity in some *B. oleracea* types.

As the seeds ripen, the plants start to dry out and the orange-brown colour of the plant is the best sign of this. The approaching maturity of the majority of seeds on a plant can also be confirmed by opening a sample of the oldest pods that will be the first to become brown. The ripening seeds will also be relatively firm in response to pressing with finger and thumb.

Many seed producers prefer to cut the ripening material by hand and place in windrows or on sheets to continue drying before extracting the seed with stationary threshers. Direct combining can only be done in dry conditions and care must be taken to minimize the loss from shattering. Figure 9.2 shows a cut brassica seed crop drying in windrows.

The effects of seed position on the plant, harvest date and drying conditions on seed yield and subsequent performance of cabbage cv. Myatt's Offenham from a crop produced on a seed to seed basis were investigated by Gray *et al.* (1985c). They found that the seeds having the most uniform, rapid and highest germination and emergence were obtained from plants cut at 30% seed moisture content.

Threshing

The seeds of *Brassica* spp. crack very easily and it is therefore important to use a relatively slow cylinder speed, not normally exceeding 700 rpm, although faster speeds may be required if the material is not very brittle.

Split seeds can be separated out from a seed lot by a spiral separator.

Fig. 9.2. *Brassica* seed crop drying in windrows.

Seed yield and 1000 grain weight

The reported seed yields vary from one area to another and also depend on the cultivar; the following figures are a guide:

- Cabbage: yield 700 kg/ha; 1000 grain weight 3.3 g.
- Cauliflower: yield 400 kg/ha; 1000 grain weight 2.8 g.
- Brussels sprouts: yield 600 kg/ha; 1000 grain weight 2.8 g.
- Kohlrabi: yield 700 kg/ha; 1000 grain weight 3.2 g.

Pathogens

The main seed-borne pathogens are listed in Table 9.1.

Basic seed production

Spring cabbage
For open-pollinated cultivars 3 years are required from sowing the stock seed until harvesting the basic seed. A further 2 years are required before the morphological characters of the seed product can be verified in field trials and the basic seed used for further multiplication to produce further generations, e.g. certified seed. The following scheme has been used in the UK for the production of basic seed of open-pollinated cultivars.

Table 9.1. The main seed-borne pathogens of *Brassica* species; these pathogens may also be transmitted to the crop by other vectors.

Pathogens	Common names
Albugo candida	White blister
Alternaria brassicae	Grey leaf spot
Alternaria brassicicola	Black spot, wirestem
Leptosphaeria maculans	Dry rot, blackleg, black rot
Mycosphaerella brassicicola	Black ring spot
Plasmodiophora brassicae	Club root
Rhizoctonia solani	
Sclerotinia sclerotiorum	Watery soft rot, drop, white blight
Pseudomonas syringae pv. *maculicola*	Bacterial leaf spot
Xanthomonas campestris pv. *campestris*	Black rot
Turnip yellow mosaic virus	

Year 1: stock seed is sown in July and the young plants transplanted into the field in September.

Year 2: during May plants are selected for optimum cultivar characters and quality. The growing point and heart are removed from the selected plants that are either left *in situ* or transferred to greenhouses (polythene tunnels have proved very suitable).

Year 3: the selected plants flower during late spring and early summer and pollinating insects are supplied if in greenhouses or other structures. Ripe seed is harvested during August. After threshing, cleaning and short-term storage the seed is either used as basic stock for further seed production or stored for a longer period while samples are grown on for field assessment, in which case samples are sown during year 4 and the sample, or progeny plots, assessed during year 5. A year can be gained by allowing the selected plants to flower in the second year by not removing their growing points.

Cauliflower

There are several techniques that are used to produce basic seed from selected plants of open-pollinated cultivars. In all cases the mother plants are grown in their normal season and final selections are made after confirmation of plant characters and curd quality. When environmental conditions in the field are expected to remain favourable for further flower development, anthesis and seed maturity, the selected mother plants can be left *in situ*. This, for example, is the normal practice in northern Europe and North America for early summer cauliflowers. However, in the case of cultivars with curds maturing in the late summer, autumn and winter the selected plants are unlikely to survive if left *in situ*.

Careful lifting of the selected plants and transferring them to a greenhouse is not always successful and techniques to produce seed from propagules of selected plants have therefore been developed.

Watts and George (1963) reviewed the range of vegetative propagation methods used and also described a method of grafting pieces of selected curd on to specially produced cauliflower rootstocks. This method was further developed by Crisp and Lewthwaite (1974) who found that a suspension of 50% a.i. benomyl applied in 500 ml water to each pot of grafted plant, with autumn cauliflower used as a rootstock, greatly improved the success rate and seed yield. Other methods of vegetative propagation include removing the stumps of selected plants to greenhouses and rooting the stem shoots that subsequently develop (Anonymous, 1980b).

The adoption of vegetative propagation techniques for cauliflower seed production has led to the establishment of clones, but there has been a build-up of viruses in the clonal material as a result of successive vegetative propagation. Turnip mosaic virus and cauliflower mosaic virus were found to be present in many of the cauliflower clones tested by Walkey *et al.* (1974) who described a method for producing virus-free cauliflowers by tissue culture.

SELECTION AGAINST BRACTING   A cauliflower curd consists of many compressed and branched peduncles with thousands of pre-floral meristems (Crisp *et al.*, 1975a). When the bracts grow through the curd surface the market value of the curd is reduced. Plants are selected against bracting during roguing or selection operations. However, because the incidence of bracting in any given cultivar is influenced by the environment, the same cultivars may display different degrees of the bracting defect in different years and production areas. Crisp *et al.* (1975a) described a two-tier system for selecting on macroscopic characters in the field, followed by approximately 4 weeks of aseptic culture during which different selections can be graded according to size of bracts.

SELECTION AGAINST UNDESIRABLE CURD COLOURS   Although the generally accepted cauliflower curd colour is white or creamy white, individual plants occur in some cultivars that have a degree of pigmentation resulting from anthocyanin, carotenoids or chlorophyll. There are also some cultivars that have curd pigmentation as a cultivar character. Coloured off-type curds are normally rogued out during roguing when the curd quality is observed. Crisp *et al.* (1975b) described a technique for improving the selection against the purple-colour defect. It involves aseptically culturing pieces of curd by the method described by Crisp and Walkey (1974) and visually inspecting the plant material for intensity of coloration after 5 weeks.

The application of tissue culture for the selection of cauliflower curd quality and its success has been discussed by Crisp and Gray (1979).

Brussels sprout

Plants selected for basic seed of open-pollinated cultivars are either left *in situ* in the field or are transplanted into cages or greenhouses, in which case it is necessary to mass pollinate them. Blowflies have been found to be the most efficient natural pollinating agent (Faulkner, 1962). Work by Johnson (1958) showed that progeny testing of open-pollinated Brussels sprout cultivars can contribute very significantly to the quality of the marketable product.

Pak Choi, Chinese Mustard: *Brassica campestris* L. ssp. *chinensis* Jusl.; Pe-tsai, Chinese Cabbage: *Brassica campestris* L. ssp. *pekinensis* (Lour) Rups.

Cultivar description of Chinese cabbage

- There are open-pollinated and hybrid cultivars.
- Plant: height short, medium or tall.
- Outer leaf: attitude; size; shape. Surface texture shape obovate, broad obovate, broad obovate to broad elliptic, colour, intensity and other salient features.
- Head at market maturity: height, width, colour.
- Time from maturity to bolting.
- Resistance to specific pests and pathogens.

(A detailed Test Guideline 105 is obtainable from UPOV, see Appendix 2.)

Soil pH and nutrition

The seed crop requires a soil pH of 6.0–7.5, and if the available calcium status is doubtful, then it should be checked and corrected with liming materials during soil preparations. The N:P:K ratio applied during site preparation is 2:1:1. Soils that have a low boron status should be corrected either by the use of boronated fertilizers or an application of borax during preparation. A supplementary top dressing of a nitrogenous fertilizer is applied at the start of anthesis especially when leaching has been significant during the vegetative stage. However, too much nitrogen will predispose the seed crop to lodging.

Irrigation

The most suitable seed production areas are where there is a relatively dry climate; this ensures that the pathological problems associated with the humid tropics are avoided or at least minimized. Furrow irrigation methods are preferable to overhead systems, especially in the more humid areas.

Plant production, sowing and spacing

The seed crop is produced by either the head-to-seed or seed-to-seed method. Detailed descriptions of these methods, including practical points with special reference to the tropics have been described by Opeña *et al.* (1988). The head-to-seed method is mainly used for the production of basic or breeders seed stocks. The seed-to-seed method is used for the production of the final seed category.

When F1 hybrid seed is being produced, the ratio of female-to-male parent is usually 1:1 or 2:1 (Takahashi, 1987).

Seed yield and 1000 grain weight

The typical seed yield for an open-pollinated cultivar is 500 kg/ha.
The 1000 grain weight is approximately 3.0 g.

Radish: *Raphanus sativus* L.

Banga (1976) discussed the history and development of the four present-day cultivated radish types, which are:

1. *Raphanus sativus* L. var. *radicula*, cultivated for its swollen hypocotyl as a field and protected crop, mainly in temperate areas of the world, although they are also cultivated in other areas.
2. *Raphanus sativus* L. var. *niger*, the larger rooted type, cultivated mainly in Asia but still locally important in Germany.
3. *Raphanus sativus* L. var. *mougri*, which has a relatively insignificant root but is cultivated as a vegetable in South-east Asia for its edible foliage and relatively long seed pods that are eaten raw, cooked or pickled.
4. *Raphanus sativus* L. var. *oleifera*, the type cultivated as a fodder crop, especially in northern Europe.

All these types described above readily cross-pollinate with each other and also with the four wild *Raphanus* species, *R. raphanistrum, R. maritimus, R. landra* and *R. rostatus*. The occasional purple-rooted off-types found in seed stocks are either a result of cross-pollination between cultivated and wild species or an admixture of seed derived from wild types that were not rogued or weeded out during seed production.

Cultivar description of radish

- There are some hybrid cultivars of which the Japanese radish ('Mooli') is prominent.
- Diploid or tetraploid.
- Seedling: anthocyanin absent or present.
- Cotyledon: size.
- Leaf: attitude, pose, length. Leaf blade features.
- Radish at market maturity: shape, colour, bicolour.
- Radish hypocotyls at market maturity: thickness, width, shape, colour, bicolour.

- Flower; petal colour at start of anthesis.
- Time of market maturity.
- Resistance to specific pests and pathogens.

(A detailed test guideline (064) is obtainable from UPOV; see Appendix 2.)

Cultivar descriptions for the 'black' radish, *Raphanus sativus* L. var. *niger* follow a similar outline to the above, but there are some variations. (A detailed test guideline (063) is obtainable from UPOV; see Appendix 2.)

A classification of forcing radish was made by Watts and George (1958), later work by George and Evans (1981) studied the morphology of 55 cultivars of radish recommended for protected cropping and classified them into 25 classes.

Soil pH and nutrition

Radish crops tolerate slightly acid conditions and a soil pH between 5.5 and 6.8 is suitable. The general N:P:K fertilizer application during site preparation is 1:3:4, although a lower amount of potassium is applied where there are satisfactory nutrient residues from previous crops. Excessive amounts of nitrogen will delay the start of anthesis and also predispose the seed crop to lodging.

Seed production systems

Both root-to-seed and seed-to-seed systems are used. The root-to-seed system is used for the biennial types, especially in Europe, where the roots are lifted in the late autumn, the tops taken off and the radishes are stored, usually in clamps, during the winter. It is also the method used for stock seed production of the annual types but in this case the material is replanted immediately after selection. In some areas of the world, especially in Asia, up to half of each steckling's root is removed before replanting. Work by Jandial *et al.* (1997) has suggested that for seed production of the Japanese white cultivars this results in a higher seed yield.

The seed-to-seed system is used for final multiplication stages where inspections of the mature root are not considered necessary and is normally used only for spring sown seed crops unless the cultivar has a vernalization requirement.

Sowing rates and spacing

Sowing rates of up to 6 kg/ha are used, the seeds are drilled in rows from 50 to 90 cm apart to give a final plant distance within the rows or approximately 5–15 cm.

For hybrid seed production the ratio of female-to-male parent rows is usually 1:1 (Takahashi, 1987).

Flowering

According to Banga (1976) there has been selection for adaptation to different day-lengths and types of seasons. The cultivars developed for early spring fresh market production have an annual habit, and material without a vernalization or specific day-length requirement has been developed from both *Raphanus sativus* var. *radicula* and *Raphanus sativus* var. *niger*. However, these types flower earlier when grown in long days.

Selection of types suitable for crop production in summer, autumn or winter has resulted in cultivars that are biennials with a vernalization requirement. The removal of early bolting plants from late spring and summer cultivars has probably resulted in the decreased sensitivity to day-length (Banga, 1976).

The flowers are cross-pollinated by bees and other insects.

Isolation

The recommended isolation distance is 1000 m, although the seed regulations of some authorities only require a minimum distance of 200 m between commercial seed stocks of similar cultivars.

Roguing stages (seed-to-seed system)

1. At market maturity stage of radish. Root: relative size, shape, colour, proportions of each colour on bicoloured cultivars, solidity.
2. At stem elongation remove wild radish types. Check that the remaining plants are true to type for foliage and stem characters.
3. At flower bud and very early at start of anthesis. As described for stage 2.
4. Flower colour.

Basic seed production

The root-to-seed system is used for basic seed production. When selecting plants according to their external morphology, care must be taken to ensure that those with pithy roots are rejected. Traditionally, selected roots were examined for internal solidity by removing a small wedge of tissue with a knife, but Watts (1960) described an immersion technique that is very suitable for screening roots for solidity at or after normal market maturity and can be used to increase the length of time that they remain solid. After selection the plant's leaves are twisted off (leaving the growing point undamaged) and the radish roots put in a bucket of water. Those roots with a degree of pithiness float and are discarded. The solid roots sink and are retained. The selected solid roots are then planted up and grown on for seed production.

Harvesting

When the seeds are nearing maturity the seed pods turn from green to brown, lose their fleshy appearance and become parchment-like. The pods do not shatter very readily and it is therefore better to harvest under very dry conditions if combined. The dry pods are then relatively brittle and seed extraction is easier. Combines with a roller attachment on some types of bean threshers are most suitable. The rollers are adjusted so as to crack (but not crush) the pods and unthreshed pods are returned to the threshing cylinder.

A standard swather is used if the seed crop is cut before threshing, and the material is left to dry further before passing it through a thresher or combine. The cylinder speed should be reduced to 500–600 rpm.

Seed yield and 1000 grain weight

A seed yield of approximately 1000 kg/ha is satisfactory, although yields of 1500–2000 kg/ha are achieved by some seed producers.

The 1000 grain weight of radish is approximately 10 gm.

Pathogens

The main seed-borne pathogens of *Raphanus sativus* with common names of the diseases they cause are listed in Table 9.2.

Table 9.2. The main seed-borne pathogens of *Raphanus sativus*; these pathogens may also be transmitted to the crop by other vectors.

Pathogens	Common names
Alternaria alternata	
Alternaria brassicae	Grey leaf spot
Alternaria brassicicola	Black leaf spot
Alternaria raphani syn. *A. matthiolae*	Leaf spot
Colletotrichum higginsianum	Anthracnose, leaf spot
Gibberella avenacea	Root and stem rot
Leptosphaeria maculans	Black leg
Rhizoctonia solani	Damping-off, canker
Xanthomonas campestris pv. *raphani*	Bacterial spot
Radish yellow edge virus	
Tobacco streak virus	
Turnip mosaic virus	

Swede, Rutabaga: *Brassica napus* L. var. *napiobrassica* (L.) Rchb.

The cultivated swede originated in Europe; it is commonly known as rutabaga in North America. In addition to its culinary value as a winter root crop it is an important stock feed crop in northern Europe. The evolution and development of the modern swedes, which are closely related to swede oil rapes and swede fodder rapes, was outlined by McNaughton (1976a).

Cultivar description of swede

Season of maturity, suitability for winter storage

- Root shape: external root colour, violet-purple, bronze, yellow or white; colour of root crown.
- Relative height of 'neck' (which is a tuberized epicotyl base).
- Resistance to specific pests and pathogens, especially *Erysiphe cruciferarum* (powdery mildew) and *Plasmodiophora brassicae* (clubroot).

(A detailed test guideline (089) is obtainable from UPOV; see Appendix 2.)

The agronomy and seed production techniques used for swede and turnip are similar; they are therefore dealt with together in the turnip section below.

Turnip (True Turnip, Vegetable Turnip): *Brassica campestris* L. ssp. *rapifera* Metz.

Turnips that originate in Europe are widely cultivated throughout the world for their edible swollen roots and leaves. However, in the tropics the crop is generally confined to the higher altitudes. The evolution and development of the turnip was discussed by McNaughton (1976b).

Cultivar description of turnip

- There are open-pollinated and hybrid cultivars.
- Diloid or tetraploid; season of maturity, suitability for early production and protected cropping, suitability for storage.
- Cotyledon: length short, medium or long; width narrow, medium or wide.
- First leaf: amount of hairiness on margin.
- Leaf: shape, colour, pose and relative size.
- Root: shape flat, globe or long globe. Colour, skin and flesh, white, purple or yellow. Pigmentation of root crown.
- Resistance to specific pests and pathogens, including *Plasmodiophora brassicae* (clubroot).

(A detailed test guideline (037) is obtainable from UPOV; see Appendix 2.)

Soil pH and nutrition for swedes and turnips

Swedes and turnips require a soil pH of 5.5 to 6.8; soils which have a pH below the lower end of this range must receive adequate amounts of liming materials during their preparation prior to sowing.

The general N:P:K fertilizer requirements applied during seedbed preparation are in the ratio of 1:2:2 or 1:1:1; the lower nitrogen ratio is used for turnips unless the soil's nitrogen status is low.

Both crops are susceptible to boron deficiency that causes 'brown heart' of their roots. Boronated fertilizers are used where the soils are known to be relatively low in this micronutrient.

Supplementary nitrogen is usually applied as a top dressing in the spring, especially in areas or seasons in which rainfall results in a high rate of leaching. However, supplementary nitrogen dressings must be carefully monitored as excessive nitrogenous fertilizers increase the incidence of lodging when the seed crop is maturing.

Plant production for swedes and turnips

Although seed-to-seed and root-to-seed systems are used for both crops, a root-to-seed system is generally used for basic seed production. The production of swede 'stecklings' for transplanting was widespread in northern Europe but has been largely replaced by the seed-to-seed method.

When seed is direct drilled, 2–4 kg/ha are sown in rows 50–90 cm apart. The crop is thinned to approximately 4–5 cm between plants within the rows. When a crop is to be transplanted, a sowing rate of 3–4 kg/ha will provide sufficient transplants ('stecklings') for 6–10 ha. In areas that are subject to severe winter frosts, the overwintering plants are protected with straw or other suitable materials. Both crop species are sown in late summer for seed production in the seed-to-seed system, but turnips are generally less hardy than swedes and early spring sowings are made in areas where the crop would not be expected to survive the winter.

For hybrid seed production of turnip the ratio of female-to-male parent is usually 1:1 (Takahashi, 1987).

Topping
Both crop species are 'topped' to encourage the development and growth of secondary inflorescences from the main flowering shoots. This 'topping' also reduces the overall height of the crop and the possibility of lodging at a later stage. The removal of the growing points also reduces the seed crop's range of maturity period. The top 10 cm of the terminal shoots are removed when the flowering shoots are between 30 and 40 cm high.

Flowering, pollination and isolation for swedes and turnips

Swedes and turnips are both biennials, and the overwintering plants are vernalized. There is a wide range of vernalization requirements among turnip cultivars, and although many of those that are of agricultural importance will flower in their first year, especially from a spring sowing, most of the vegetable turnip cultivars will not.

Both crop species are mainly insect-pollinated although some wind pollination also occurs. The vegetable cultivars of swede readily cross-pollinate with cultivars of agricultural swede, swede oil rape, swede fodder rape, kale rape and the turnip group. The vegetable turnip cultivars cross-pollinate with agricultural turnips, turnip oil rape, turnip fodder rape and the swede types listed above. The recommended isolation distances for swedes and turnips are 1000 m.

Roguing stages for swedes and turnips

Seed to seed

1. Early vegetative stage, before swelling of the roots: check leaf type, colour and relative height.
2. Start of anthesis: check that the flower colour and size is according to type.

Root to seed

1. Early vegetative stage, before swelling of the roots: check leaf type, colour and relative height.
2. When roots are lifted for storing (or replanting, depending on local winter climate and custom): check that root shape, relative size, colour of root and shoulder are according to type.
3. When replanted crop reaches anthesis: check that flower colour and flower size are according to type.

Basic seed production for swedes and turnips

Only the root-to-seed system is used for the production of basic seed. Roots for selection should be grown to normal market maturity by the autumn of their first year. They are then selected on root characters as described above. The early maturing turnip cultivars are sown later for the early market crop, otherwise the roots are too large by the late summer and autumn and have a tendency to rot during storage.

The selected roots are stored in frost-proof conditions in moist peat or sand and replanted in the late winter when no further severe frosts are expected. There is an increased tendency to replant the selected roots in polythene tunnels at the end of the winter.

Harvesting for swedes and turnips

Seed crops of both swede and turnip have a tendency to shatter readily and therefore careful cutting is required to minimize unnecessary crop loss during harvesting. Generally the best indication that the bulk of seeds are near maturity is that the plant haulm turns from green-brown to parchment colour. The crop is cut with a swather and left in windrows until the seeds are mature and separate easily from their pods. The material is then either picked up from the windrows by combine or fed into a thresher. Material produced on a small scale or in the protection of polythene tunnels is hand-harvested.

Seed yield and 1000 grain weight for swedes and turnips

A satisfactory yield for both crop species is 1500 kg/ha, although yields of up to 2500 kg/ha are achieved in the best seed production areas of the USA.

The 1000 grain weight of turnip is approximately 4.3 g, while the 1000 grain weight of swede is approximately 3.3 g.

Watercress: *Rorippa nasturtium-aquaticum* (L.) Hayek, syn. *Nasturtium officinale* R. Br.

Origins and types

The early history of watercress, which originated in northern Europe, was discussed by Howard (1976). It is now cultivated in Europe, Asia, North America, South Africa and Australasia. Watercress is grown commercially as an aquatic crop although plants will thrive in a soil substrate without aquatic conditions. The leaves and shoots are eaten as a salad, although in some areas they are cooked.

The species cultivated widely in the UK and to some extent elsewhere in Europe is *Rorippa nasturtium-aquaticum* (L.) Hayek commonly called 'green-cress'. It has largely replaced *Rorippa microphylla* (commonly called 'brown-cress'), which is sterile and was therefore propagated vegetatively. In addition to its more attractive colour, the 'green-cress' has other advantages. It is less susceptible to crook root disease (Spencer and Glasscock, 1953), and virus-free stocks can be raised from seed (Tomlinson, 1957). There has therefore been a very keen interest in developing seed production techniques for this crop.

There are no cultivars although many growers have produced and maintained their own selections, clones or 'strains'.

Agronomy and crop production

The crop is maintained in the same way as the market crop except that beds for seed production are inspected and rogued especially for off-types and the presence of viruses.

Flowering and pollination

Work by Bleasdale (1964) showed that 'green-cress' is a long-day plant that flowers in response to the increasing day-length of spring and summer in northern Europe, and it was suggested that late flowering strains could be selected. Although late-flowering strains would prolong the market season of the vegetative shoots, the length of season for seed production would be reduced. With this in mind some seed producers have developed techniques of either transferring plants from aquatic beds to a soil substrate in greenhouses or putting temporary plastic structures, such as inflated 'bubble-greenhouses' over the watercress beds.

Watercress flowers can be both cross- or self-pollinated. Several species of insects, including *Diptera* species, are pollinating agents.

Isolation and roguing

There are no specific recommendations in the literature on isolation distances for seed production of this crop, but a minimum distance of 1000 m would be appropriate.

The beds of plants for seed production are generally rogued during the autumn, winter and spring prior to flowering. Plants showing symptoms of virus infection, any other pathogens, pale-green leaf colour and early flowering are discarded. The most important virus of watercress is turnip mosaic virus, although it is not seed-transmitted. Smaller, weaker plants that are otherwise healthy are also discarded in order to maintain or increase the vegetative productivity of the strain.

Harvesting and 1000 grain weight

Watercress, like other *Cruciferae* seed crops, has a tendency to shatter. The flowering stems are cut and placed in paper-lined trays to dry as soon as the majority of seeds are mature. The seeds are a pale-yellow colour at this stage and later turn a darker brown as they ripen and dry further. The seeds are usually extracted by hand rubbing the plant material and then passing through a series of seed sieves to remove pieces of plant debris.

The 1000 grain weight is approximately 2.8 g.

Further Reading

Brassicas including Chinese cabbage

Dixon, G.R. (2006) *Vegetable Brassicas and Related Crucifers.* CAB International, Wallingford, UK.
Herklots, G.A.C. (1972) *Vegetables in South-East Asia.* George Allen and Unwin Ltd, London, pp. 182–224.
Shinohara, S. (1984) *Vegetable Seed Production Technology of Japan, Elucidated with Respective Variety Development Histories, Particulars,* Volume 1. Tokyo, Japan, pp. 1–122.

Chinese cabbage

Opena, R.T., Kuo, C.G. and Yoon, J.Y. (1988) *Breeding and Seed Production of Chinese Cabbage in the Tropics and Subtropics*. Technical Bulletin No.17, Asian Vegetable Research and Development Center Tropical Information Service, Taiwan.

Radish

Shinohara, S. (1984) *Vegetable Seed Production Technology of Japan, Elucidated with Respective Variety Development Histories, Particulars*, Volume 1. Tokyo, Japan, pp. 195–237.

Swedes

Bowring, J.D.C. and Day, M.J. (1977) Variety maintenance for swedes and kale. *Journal of the National Institute of Agricultural Botany* 14, 312–320.

10 *Cucurbitaceae*

The cultivated genera of this family are widely distributed in the world. They are referred to as 'vine' crops in some areas, especially in North America. All the species are susceptible to frost, and while most are important outdoor crops, e.g. watermelon in the southern states of the USA and many tropical and arid areas, some, e.g. cucumber, have also become important protected crops, especially in northern Europe and parts of Asia.

Some of the crops, e.g. watermelon and melon, are not regarded as vegetables but as 'fruits' by some authorities because they are eaten as dessert. However, because of the production methods of the market and seed crops they are generally considered along with the vegetable crops.

The majority of the *Cucurbitaceae* species have unisexual flowers borne on monoecious plants, *Telfairia occidentalis* Hook. f. (commonly known as 'fluted pumpkin' or 'telfairia nut') has unisexual flowers on separate plants. The control of sex expression in the cucurbits has been discussed by Robinson and Decker-Walters (1997).

There are several interesting problems in the production of cucurbit seed, including the extraction of seed from 'wet' fruit, the cross-compatibility of some of the different crops or cultivar types within the same genera and the positive identification of plants for seed production based on immature fruit characters. This last point, coupled with the fact that the majority of cultivated cucurbits are predominantly cross-fertilized, requires careful attention to isolation and roguing especially for the production of basic seed.

Citrullus lanatus (Thunb.) Mansf. Syn. *Citrullus vulgaris* Schrad.	Watermelon, egusi
Cucumis melo L.	Cantaloupe, melon, sweet melon
Cucumis sativus L.	Cucumber
Cucurbita maxima Duch. ex Lam.	Pumpkin, winter squash, Chinese pumpkin, crookneck squash

Cucurbita moschata (Duch. ex Lam.) Pumpkin, winter squash
 Duch. ex Poir
Cucurbita mixta Pang.   Pumpkin, winter squash
Cucurbita pepo L.   Marrow, vegetable marrow, courgette

The current thinking on the origins, nomenclature, descriptions and the distinguishing characters of the *Cucurbita* species has been outlined by Robinson and Decker-Walters (1997).

Watermelon: *Citrullus lanatus* (Thunb.) Mansf.

This crop is widely grown throughout the tropics, subtropics and arid regions of the world. It originates from Africa but there are references to the very early cultivation of the species elsewhere in the world (Hedrick, 1972). In some parts of Africa, local cultivars with relatively bitter fruits are cultivated for their seeds which are roasted and eaten; these are referred to as 'egusi', but production for this purpose is not generally included as part of seed production.

Watermelons have been further selected and improved by seedsmen and plant breeders, especially in the USA and parts of Asia, where a very wide range of cultivars is maintained. Some of the cultivars originally bred in the USA, e.g. 'Charleston Gray', are cultivated in many areas of the world.

The watermelon, a vigorous annual which covers a large area of ground with its sprawling stems, can survive relatively dry conditions because it roots deeply. For this reason it has become established as an important crop in many developing countries, especially where arid conditions prevail.

Cultivar description of watermelon

- There are open-pollinated and hybrid cultivars.
- Ploidy: diploid, triploid, tetraploid.
- Season of cultivation: including suitability for intensive and/or extensive protected cropping.
- Plant vigour: bush, vine, extensive vine.
- Foliage: leaf protection against sunscald.
- Fruit: external shape; round, oval or long relative size and weight at market maturity rind colour, bicolour, colour pattern, striped internal; thickness of rind, flesh colour, intensity of flesh colour, especially towards the centre of the fruit; absence of central cavity.
- Seed: external colour when mature, striped or single colour, relative size.
- Resistance to specific pests, pathogens or other disorders, e.g. *Fusarium* sp. *Didymella* (gummy stem blight) and *Colletotrichum* sp. (anthracnose) and sunscald.

(A detailed test guideline (142) is obtainable from UPOV; see Appendix 2.)

Soil pH and nutrition

The watermelon plant tolerates soils with a pH from 5.0 to 6.8, although most sites used for commercial seed production have a pH between 6.0 and 6.8. Many small-scale watermelon producers apply a bulky organic manure during soil preparations at a rate of up to 25 t/ha if locally available, but this is often not possible for large-scale production.

In the absence of bulky organic manures the base fertilizer ratio dressing applied should be 1:1:1 of N:P:K according to the nutrient status of the soil. Higher nitrogen ratios than this should be avoided, otherwise there is an excessive vegetative growth of the plants.

Agronomy

In most areas of the world where watermelons are grown for seed production, seeds are sown direct into the field in preference to raising plants in nurseries.

The plants are either grown on 'mounds', flat ridges or on the flat. The system adopted locally depends on the irrigation system to be used, and to some extent on custom. Mounds or ridges are used in conjunction with a furrow or similar irrigation system. The seeds are spot sown, usually two to three seeds per station, 90–120 cm apart in the row, with 120–180 cm between the rows; 1–3 kg of seed are sufficient to sow 1 ha. A higher seeding rate is required when the seeds are drilled.

The seed is slow to germinate early in the season if soil temperatures are low. The plants are very frost-sensitive and local custom generally avoids a sowing date before the incidence of frosts has passed.

The seedlings are thinned out to their final stand when the first true leaves are showing. Care must be taken with crops scheduled for seed production to ensure that only one seedling is left per station, this is especially important for the production of basic seed.

The frequency of irrigation will depend on the soil type and climate, but because watermelon plants develop a deep and extensive root system applications can be kept to a minimum; in dry regions, sufficient irrigation should be applied prior to sowing to restore the soil to field capacity.

Flowering

Watermelon is day-neutral and there are therefore no problems in flower initiation. However, plant and fruit development are poor when ambient temperatures are less than 25 °C.

Pollination

Watermelon flowers are insect-pollinated, mainly by honeybees. The plants are self-compatible, but because flowers are unisexual a high percentage of cross-pollination occurs.

It is a normal practice in the USA to place colonies of hive bees on the perimeter of watermelon fields. This is to help increase seed yield and it is also claimed that by supplying a high population of pollinating insects adjacent to, or within, the seed production plot or field, the incidence of cross-pollination with other fields of watermelon which may be a different seed category or cultivar is minimized.

Isolation

The minimum recommended isolation distance for watermelons intended for seed production is 1000 m. The recommended distances for stock seed are at least 1500 m. It is important to ensure that seed crops are also isolated from market and private garden crops.

Roguing stages

1. Before flowering: check vegetative characters.
2. At early flowering: check trueness to type of developing fruit.
3. Fruit developing: as for stage 2.
4. Marketable fruit: check fruit characters.

Plants infected with any of the seed transmitted pathogens should be removed from the seed crop as soon as identified, at all stages.

Harvesting

Sufficient time must be allowed for the seed to reach maturity, which is at least a week later than the optimum stage for harvesting the fruit for the fresh market. The harvesting stage for seed production can usually be confirmed when the tendrils on the shoot bearing the individual fruit have withered. Another sign that the fruit has reached maturity is the colour change from green/white to pale yellow on the underside of the fruit (i.e. the surface which has been resting on the soil).

The method of collecting fruit for seed extraction depends on the scale of operation. In the USA, where fields for production of watermelon seed are often at least 10–20 ha, the whole operation is mechanized. In countries which use a high percentage of hand labour, the entire operation may be done by hand, especially in smaller areas for stock seed or commercial seed production.

The cucurbit-harvesting machines (usually referred to as 'vine harvesters') which have been developed in the USA are either self-propelled or towed by a tractor. The self-propelled machines are capable of picking up the fruits from the plants, but this is done when the entire crop has reached maturity, as it is a once-over harvest which collects all the fruit and destroys the plants. A tractor-towed vine harvester is illustrated in Fig. 10.1.

Hand-harvesting methods are based on the selective cutting of mature fruits that are put directly into a cucurbit seed extractor. This machine is towed

Fig. 10.1. Mechanized cucurbit harvesting and seed extraction.

through the field at a speed compatible with the rate of cutting, according to the number of workers available.

An alternative system is to cut the mature fruits and place them in windrows or heaps either to await the seed extractor which goes through the field later, or they are collected for immediate transportation to a central area where the extraction is done.

Seed extraction

The seeds of the watermelon fruit, unlike most cucurbits, are distributed throughout the central area of the fruit pulp and are not in a central fruit cavity. Therefore, hand-extraction methods depend on fruit maceration rather than a scooping process. The macerated pulp containing the seeds is washed by running water into the screen; this operation separates the pieces of rind and coarse material from the seeds and fine pulp. The seeds pass through to a finer meshed screen which retains them. The better the initial maceration of pulp, the more efficient is the separation by screens, resulting in a cleaner final seed sample.

Fermentation is not normally used in the extraction or cleaning of watermelon seed because the seed easily discolours and potential germination is reduced by fermentation.

Drying

The drying of watermelon seed must commence as soon as the extraction is completed. Large rotary driers are used by specialist watermelon seed producers, especially for the preliminary drying period. The control of air temperature is not very accurate with rotary driers and many seed producers prefer the batch driers with a rotary paddle The initial air temperature for drying is usually 38–41 °C and as the pieces of fruit and rind debris dry off (i.e. there is no obvious free moisture when a piece of pulp is pressed between the fingers), the temperature is reduced to 32–35 °C. The drying is continued until the seed moisture content does not exceed 10%. They are dried to a moisture content of 6% for vapour-proof storage.

Seed yield and 1000 grain weight

The average seed yield of most watermelon cultivars under good conditions is 400 kg/ha. The popular cultivar 'Charleston Gray' is a relatively low seed yielder, and an average yield is only 250 kg/ha. Under the relatively poor cropping conditions found in some tropical countries, seed yields of the better-yielding cultivars are as low as 100 kg/ha, with cv. 'Charleston Gray' yielding even less. A relatively short growing season also affects seed yield adversely. A lower yield of fruit is achieved when the fruit is left on the plant for seed production than if fruits are promptly harvested as they reach market maturity.

The 1000 grain weight depends on the cultivar, but the approximate weight is 113 g.

Stock seed production

Foundation seed of open-pollinated cultivars is maintained by selfing single plant selections which are grown in isolation. After harvesting the seed separately from each selected plant, the individual seed lots are progeny-tested and the approved material is bulked for further multiplication.

Pathogens

The main seed-borne pathogens of *Citrullus* are listed in Table 10.1.

Table 10.1. The main seed-borne pathogens of *Citrullus* species; these pathogens may also be transmitted to the crop by other vectors.

Pathogens	Common names
Colletotrichum sp.	
Didymella bryoniae	Gummy stem blight, black rot
Fusarium oxysporum f. sp. *niveum*	Wilt
Glomerella lagenaria	Anthracnose
Pseudomonas pseudoalcaligenes ssp. *citrulli*	
Squash mosaic virus	

Cucumber: *Cucumis sativus* L.

This species is widely cultivated for the production of its edible fruits, which are used in salads. Small fruited types (often referred to as gherkins) are also used in pickles. In parts of Asia young cucumber leaves are eaten raw or cooked similarly to spinach. Cucumbers are cultivated as a field crop in most areas of the world under frost-free conditions, but are also an important heated greenhouse or plastic structure crop in some temperate areas such as northern Europe. The early protected crop production has now become an important development in many areas, including the Middle East and parts of Asia. The traditional greenhouse cultivars develop parthenocarpic fruits if not pollinated, and this character is now incorporated into many of the outdoor types.

Origins and types

Cucumber cultivars are generally classified into five groups:

1. Field cucumbers, with prominent black or white spines.
2. The greenhouse or forcing type (often referred to as 'English cucumber'). These have spineless fruit which can be produced parthenocarpically; there are also short fruited cultivars in this group.
3. The Sikkim cultivars (originating from India) with reddish-orange fruits.
4. The small fruited cultivars frequently known as 'gherkins' for pickling.
5. The 'apple' fruited cultivars which have ovoid- to spherical-shaped fruit, from diverse areas of the world including the USA and the Far East.

Cultivar description of cucumber

- There are open-pollinated and hybrid cultivars.
- Use: salad, pickling; greenhouse, protected or outdoor production.
- Fruit characters: relative size and shape, shape of basal end.
  - Rind colour at market maturity, and at seed maturity.
  - Spines: degree of spininess, spine colour (black or white) and relative size of 'warts' at bases of spines.
  - Ability of fruit to retain characteristic shape if picking for fresh market is delayed.
  - Freedom from bitterness.
- Vegetative characters: determinate or indeterminate.
- Resistance to:
  - Scab or gummosis (*Cladosporium cucmerinum*).
  - Powdery mildews (*Sphaerotheca fuliginea* and *Erysiphe cichoriacearum*).
  - Downy mildew (*Pseudoperonospora cubensis*).
  - Cucumber leaf blotch (*Corynespora melonis*).
  - Cucumber mosaic virus (CMV).

(A detailed test guideline (061) is obtainable from UPOV; see Appendix 2.)

Soil pH and nutrition

Cucumber plants require a soil with a pH of 6.5 or slightly above. The crop responds to soils with a relatively high organic matter content, therefore the site should receive a dressing of up to 80 t/ha of decomposed organic manure during the early stages of preparation.

A suitable base N:P:K nutrient ratio application of 1:2:2 should be applied during the final stages of seedbed preparation, but the nutrient value of any bulky organic manures already applied should be taken into account. The amount of nitrogen is increased in soils with a high phosphorus and potassium status. A higher proportion of nitrogen is also necessary where frequent irrigation is required in order to allow for leaching; in this situation the N:P:K ratio should be nearer to 2:1:1, with approximately half the nitrogen applied as a top dressing about a month after seedling emergence. Care should be taken to avoid foliar scorch from this operation.

Plant establishment

Seeds of open-pollinated cultivars are normally sown direct into the field at stations 10–12 cm apart, with up to 2 m between the rows. In arid areas where irrigation channels are necessary, the seeds are sown on flat ridges which are up to 30 cm high. The seedlings are thinned to a single plant per station as soon as they have emerged. Similar systems are used for hybrid seed production except that a closer row spacing of 50 cm is usually adopted, and there are six to eight rows of the female parent between two male parent (or pollinator) lines; this pattern is repeated across the field.

The plants are 'stopped' by pinching out the initial leader between three and five leaves; two main laterals are subsequently secured. About four to five fertilized fruits are obtained per plant in the female rows. A similar stopping system is usually adopted for open-pollinated cultivars.

Sex expression in cucumber flowers

Cucumber plants produce both male and female flowers. The monoecious cultivars normally produce these in approximately equal proportions, although there is a tendency to produce only male flowers initially.

Sex expression in cucumber is generally influenced by environment. Under long days and high light intensities male (staminate) flowers predominate, whereas under short days and low light intensities female (pistillate) flowers predominate.

Plant breeders have produced cultivars which under normal conditions bear predominantly or completely female flowers. These are usually referred to as gynaecious cultivars. The advantages of these, many of which are F1 hybrids, are that they are earlier, higher yielding and all fruits are parthenocarpic and therefore seedless. In addition, some of the F1 hybrids have resistance to specific pathogens.

There is an obvious technical problem for the seed producer who has to manipulate the plant in order to produce a higher proportion of male flowers to ensure an adequate supply of pollen. Different breeders and seed producers have developed their own techniques for producing staminate flowers on cucumbers based on experiences with individual cultivars or lines in specific environments or locations. These are based on the use of gibberellic acid or silver nitrate. Three alternative methods have been developed, these are:

1. Three applications of GA3 at 1000 ppm, sprayed on at fortnightly intervals, commencing when the plants have two leaves.
2. As above, but using GA4/7 at 50 ppm.
3. A single application of silver nitrate solution (600 mg/l) before the first flowers open.

Hybrid seed is collected only from the female parent plants; the presence of any pistillate flowers on the male plants is not a problem but it is important that staminate flowers are completely suppressed on the female parent. This is normally achieved by two applications of ethrel (250 ppm), the first when the plants show their first true leaf and the second at the fifth true leaf stage. A visual check must also be made and any male flowers on the female parents removed by hand. The development of suitable male sterile lines or the application of a satisfactory gametocide would be a useful development in the production of hybrid cucumber seed.

A further safeguard which avoids admixture of seeds from the male parents is to rotavate the male parent rows before harvesting commences. The advice and instructions of the maintenance breeder should be followed throughout the programme.

Pollination

Cucumbers are self-compatible, but are predominantly cross-pollinated. Pollination is mainly achieved by bees when the plants are grown as a field crop. When the seed is produced in greenhouses, attention should be given to ensure that adequate pollinating insects are visiting the crop unless hand pollination is being undertaken. *Cucumis sativus* is not cross-compatible with *Cucumis melo*, gourds, marrows, squash, pumpkins or watermelons.

Isolation

Cucumber seed production in fields or plots should be at least 1000 m from all other cucumber crops including those for market production. It is especially important that the different types of cucumbers have sufficient isolation during production of commercial seed. For basic seed production the isolation distance should be at least 1500 m. These recommended distances can be reduced when seed is produced in insect-proof structures.

Roguing stages and main characters to be observed

Seed crops intended for commercial category seeds are usually only examined and rogued at stages 3 and 4.

1. Before first flowers open: desirable characters; growth habit, vigour and foliage typical of the cultivar.

2. Early flowering: plant habit, foliage characters as checked in stage 1.
- Observable characters of undeveloped fruit, especially colour of spines.
- Check for any specific seed-borne pathogens present.

3. Fruit setting: as for stage 2. Also:
- Satisfactory level of productivity.
- Time of main production.
- Fruit characters, including size, shape and colour.

4. Ripe fruit: colour of ripe fruit in accordance with cultivar description, e.g. fruits green, yellow, white or orange.

Harvesting

Ideally the fruit should remain on the plant until it is fully mature. This is indicated externally by the development of the ripe rind colour characteristic of the cultivar; additionally the fruit stalk adjacent to the fruit withers when the seed is mature.

In order to confirm the external signs that the seeds are actually mature, several fruits should be cut open longitudinally and the seeds examined. Mature seeds separate easily from the interior flesh.

The mature fruit are hand-picked and put in a fruit crusher and seed extractor (as described for watermelon). Large-scale specialist producers use mechanized harvesting machines incorporating crushers and seed extractors. If the seeds are extracted by hand, the ripe fruits are cut in halves longitudinally and the seeds are scraped into a container.

The seed and juice mixture can be fermented for about a day before screening and washing the seed in suitable-sized sieves. Water troughs with riffles, as described for tomato seed washing, are used in large-scale operations.

The seed is dried as described for watermelon. After drying, the seeds are screened to remove any remaining fruit debris. Aspirated screens will remove light and immature seeds.

Seed yield and 1000 grain weight

The average seed yield under field conditions is 400 kg/ha, although yields of up to 700 kg are often reported. The yield from a single fruit depends on the cultivar and the amount of successful pollination, but estimates can be based on approximately 500 seeds per fruit. The seed yield for F1 hybrids produced in fields with a male/female population ratio of 1:4 is 300–350 kg/ha.

Table 10.2. The main seed-borne pathogens of *Cucumis* species; these pathogens may also be transmitted to the crop by other vectors.

Pathogens	Common names
Alternaria cucumerina	Leaf spot
Colletotrichum lagenarium	Anthracnose
Didymella bryoniae	Leaf spot, Black rot
Fusarium oxysporum	Fusarium
Pseudomonas syringae pv. *lachrymans*	Angular leaf spot
Cucumber green mottle mosaic virus	
Cucumber mosaic virus	

The 1000 grain weight of the smaller fruited cultivars is 25 g. The seeds of the longer fruited greenhouse cultivars have a 1000 grain weight of 33 g.

Pathogens

The main seed-borne pathogens of *Cucumis* spp. with common names of the diseases they cause are listed in Table 10.2.

Melon: *Cucumis melo* L.

The fruits of this species are eaten as a dessert and are not normally cooked, although some are used in preserves, yogurts, and canned or frozen melon balls.

Melons are cultivated throughout the tropics, subtropics and to a limited extent in parts of the temperate regions which have relatively long warm summers. Winter production in parts of Africa (e.g. the Sudan and Kenya) for export to northern Europe has increased its importance as a cash crop, and its cultivation has also increased as a cash crop in some South American countries. Melons have also been grown as protected crops in heated greenhouses in northern Europe, but their production as protected crops is now largely confined to unheated polythene tunnels in the Mediterranean region where they have become established as an important export crop.

There is a very wide diversity of fruit types and local selection in different areas of the world has led to a large number of cultivars. Pursglove (1974) divided the cultivated melons into four groups that include:

1. The European cantaloupe melon, with thick, scaly, rough rind which is sometimes grooved.
2. The musk melon, cultivated in the USA, which are smaller fruits than the cantaloupe melon, and the rind finely netted or almost smooth with shallow ribs.
3. The Casaba or winter melon with large fruits, late maturing, with good storage quality. The yellow or greenish rind is usually smooth, some cultivars striped or splashed; the flesh has a slightly musky flavour.
4. The oriental types, with elongate fruits nearer to cucumbers in shape, and often used as vegetables.

Cross-pollination between different cultivars occurs freely and there are therefore many intermediate types. Robinson and Decker-Walters (1997) give details of a classification derived from a system originally outlined by Naudin in 1882 in which he proposed ten divisions. The system described by Robinson and Decker-Walters has six groups based on horticultural uses and takes into account fruit morphology; with cantaloupe and musk melons in the 'Cantalupensis' group.

The use of isoelectric focusing in immobilized pH gradient (IEF-IPG) of melon seed protein has been reported as a useful tool for interpreting the genetic variability in melons (van den Berg and Gabillard, 1994).

Cultivar description of melon

- There are open-pollinated or hybrid cultivars.
- Season: outdoor type, suitability for greenhouse production, suitability for protected cropping, suitability for shipping, storage.
- Fruit characters: shape; ratio of length to width.
- Rind: external colour and/or colour pattern, external rind texture.
- Flesh: colour, sweetness, relative width of cavity and flesh.
- Specific pathogen and pest resistance.

(A detailed test guideline (104) is obtainable from UPOV; see Appendix 2.)

Soil pH and nutrition

All the melon types require soils with a pH of between 6.0 and 6.8. The crop has traditionally been given bulky organic manures incorporated during preparation although many areas of the world produce melons of excellent quality and high seed yields without the application of bulky organics. This is very evident in the Middle East and parts of Asia where melons are produced on reclaimed desert soils with available irrigation. The traditional applications of bulky organic manures for this crop can be replaced by using rotations of forage crops, such as alfalfa (also commonly known as lucerne; *Medicago sativa* L. and *Medicago* X *varia*) which improve soil structure.

Fertilizers applied during the final stages of seedbed preparation should contain an N:P:K ratio of 1:1.5:2, although appropriate reductions of the phosphorus and potassium levels must be made to allow for residues which are present. Where leaching is likely to be significant some of the nitrogen can be applied as a top dressing at the start of anthesis. Applications of nitrogenous fertilizers should not be given later than this stage as fruit ripening and maturity may be delayed.

Sowing and spacing

Melons are grown on the flat, on ridges or on raised beds. The choice of system usually follows the local custom which has developed according to irrigation systems used and the soil's drainage character.

Seeds are usually sown direct, two to three per station, thinning to a single plant as soon as practical. The row width used depends on the vigour of the cultivar as well as the need for irrigation channels. Rows are from 1.25 to 2.0 m apart. Distances between plants within the rows are 90 cm at the narrower row widths and down to 30 cm at the wider row widths: 2 kg of seed can be sown in 1 ha. Large-scale producers drill the seed on the flat.

The leaders and laterals are stopped when there are about four fruits per plant.

Pollination

Supplementary colonies of hive bees should be placed adjacent to fields when the natural bee population is seen to be relatively low. Plants can be hand-pollinated if small numbers of plants are grown for stock seed production, especially in greenhouses, but this is not normally economical for large-scale production outside.

Isolation

The ideal minimum isolation distance for commercial seed production is 1000 m, although some seed regulatory authorities specify distances of only 500 m. The different groups of melons within *Cucmis melo* are all cross-compatible. It is, therefore, extremely important that there is adequate isolation between commercial or garden crops grown only for fruit and crops intended primarily for seed production.

The melons in the different groups of *Cucumis melo* do not cross-pollinate with the watermelons (*Citrullus lanatus*), nor do they produce viable seeds if cross-pollinated with the cucumbers (*Cucumis sativus*).

Roguing stages and main characters to be observed

1. Before flowering: vegetative habit, characters of foliage and undeveloped fruit.
2. After flowering commences: check fruit type and morphology.
3. Fruit set: check fruit type, morphology and external characters.
4. Harvesting: check fruit type, morphology and external characters.

In addition check for seed-borne pathogens at all stages.

Harvesting

The fruit of the cantaloupe and musk melon types tend to separate from the stem at the base of the fruit as the fruit becomes fully mature. This stage of separation by formation of an abscission layer between the stem and a mature fruit is usually referred to by melon growers as 'full slip'. Large-scale melon seed producers leave the entire crop or fruit to separate in this way before passing through the field with a cucurbit seed harvester or picking up by hand and conveying the fruit in baskets to the seed extractor which moves through the field.

The fruits of the winter melons do not form an abscission layer when mature. Maturity is indicated by external rind colour change from green to yellow or yellow to white (according to the rind colour of the cultivar). In addition to the external colour change, the blossom end of the soft fruit softens and the coat becomes waxy as its aroma increases.

Melon seed is not fermented before washing to separate the seed from other fruit material. After washing the seed is dried as described for watermelon. Final separation of the seeds to be retained from the remaining fruit debris and light seed can be made by passing the seed lot through an aspirated screen cleaner.

Seed yield and 1000 grain weight

The average yield of seed is approximately 300 kg/ha, although the best melon seed production areas can achieve up to 600 kg.

The 1000 grain weight of melon seed is approximately 25 g.

Pathogens

The main seed-borne pathogens of *Cucumis melo* with common names of the diseases they cause are listed in Table 10.3.

Table 10.3. The main seed-borne pathogens of *Cucumis melo*; these pathogens may also be transmitted to the crop by other vectors.

Pathogen	Common names
Cladosporium cucumerinum	Scab
Fusarium solani f. sp. *cucurbitae*	
Fusarium sp. (possibly *F. oxysporum* f. sp. *melonis*)	Wilt
Glomerella lagenaria	Anthracnose
Macrophomina phaseolina	
Pleospora herbarum	Leaf spot
Cucumber mosaic virus	Mosaic
Melon necrotic spot virus	
Musk melon mosaic virus	
Squash mosaic virus	
Tobacco ring spot virus	

Marrows, Pumpkins and Squashes

This group of cucurbits contains a wide range of types within *Cucurbita pepo*. Robinson and Decker-Walters (1997) have discussed and described the cultivar groups considered to be within this species.

Cultivar description of vegetable marrow, pumpkin and squash, *Cucurbita pepo* L.

- There are open-pollinated and hybrid cultivars.
- Season: storage potential, use of immature fruit, and mature fruit.
- Plant habit: trailing (vine) or bush, leaf characters.
- Leaf: size small, medium or large. Leaf morphology, size, and marbling.
- Fruit: external shape and form before anthesis.
- External shape, form and colour at market maturity, bicolour and patterning.
- Internal characters: flesh colour and seed cavity.
- Resistance to specific pathogens, including *Fusarium* species.

(Detailed test guidelines for vegetable marrow and squash (119) and pumpkin (155) are obtainable from UPOV; see Appendix 2.)

Soil pH and nutrition

This group of *Cucurbita* species is moderately tolerant of acid conditions and is produced successfully on soils with a pH between 5.5 and 6.8.

While these vegetables respond to applications of up to 30 t/ha of bulky organic manures during site preparation, reasonable crops can be produced by applications of inorganic fertilizers. The optimum N:P:K ratio is 1:2:2 with appropriate deductions made for nutrients applied as bulky organic materials. Most cucurbit seed producers apply a top dressing of 1:1:1 at a rate of 60 kg/ha when the first fruits start to set. On soils with a high P and K status, top dressings of only nitrogen are used. The nitrogen top dressing is particularly important when leaching has occurred.

Sowing

There are three methods of field production: growing on the flat, on flat ridges or on mounds (sometimes referred to as 'hills'). The system adopted depends on the irrigation system and efficiency of soil drainage as this crop is not successful in waterlogged soils. The preparation of mound systems is very labour-intensive as it is difficult to mechanize the operation.

The ultimate planting density depends on the type of cucurbit and whether it has a trailing ('vine') or bush habit in its vegetative growth.

Row spacings of 90 cm to 3.5 m are used according to the vigour of the cultivar to be grown. The lower row width is used for the less vigorous and bush cultivars, whereas the wider row widths are used for the more vigorous trailing cultivars. The distance between plants within the rows is usually the same as the distance adopted between rows; although where production of the crop is fully mechanized a closer spacing between the rows is used to compensate for a wider row spacing and allows for machinery to be used during cultivations. The sowing rate is 2–4 kg/ha depending on the plant density required. It is normal practice to sow about three seeds per station and single the seedlings after emergence unless the seeds are sown with a precision drill.

Pollination

Each of the *Cucurbita* species discussed here is cross-pollinated within the species. While there is some cross-incompatibility between some pairs of these species, for seed production purposes it is safer to assume that crossing between types will take place. The reason for this assumption is that it is not always possible to confirm to which of the species a particular cultivar belongs.

Pollination is generally by bees, although some other insect species are known to pollinate *Cucurbita* flowers.

Isolation

The recommended isolation distance between seed crops within this group of species is 1000 m. It is important to avoid areas where even a few plants of these types are grown domestically for fruit production.

Production of hybrid *Cucurbita* seed

F1 hybrid cultivars have been developed by some of the seed companies specializing in seed supply of this crop. This has become especially important in the USA and seed is also produced there for export to other parts of the world.

A ratio of one male parent row to five female parent rows is generally used for hybrid seed production. An ethrel solution (250 ppm) is used to suppress the male flowers in the female plant rows. It is applied as a spray at three stages of plant development, i.e. first true leaf, third true leaf and fifth true leaf. By the time that the first flowers open at about the sixth to seventh leaf stage, all male flowers have been suppressed.

By the time that two to three fertile fruits have developed on each mother plant, the ethrel effect has gone. Further sprays would not be effective, so development of later male flowers is stopped by cutting off the plant's growing point with a knife. The recommendations of the maintenance breeder should be followed.

Roguing stages

1. Early vegetative stage: check that vegetative characters (e.g. bush or trailing type), foliage and vigour are in accordance with the cultivar, also that resistance to specific pathogens is according to the cultivar description.
2. Before first flowers open: as above, and check that undeveloped fruit characters on female flower buds are true to type.
3. First female flowers setting: check that developing fruits are true to type, and as above.
4. Fruit developing: fruit characters true to type, and resistance to specific pathogens according to cultivar description.
5. Mature fruit at harvest: fruit characters true to type.

Harvesting

All squashes, pumpkins and marrows take approximately 16 weeks from anthesis to seed maturity. By this stage the rind has hardened and usually changed colour. The green types change to a yellow-orange colour and the yellow-golden types change to a straw colour. Figure 10.2 shows a field of crookneck squash ready for harvest and seed extraction.

For large-scale production fruit may be placed in windrows ready for the mobile thresher and extractor. Alternatively they are left on the mother plant for a vine harvester to pass through the field.

The fruits of some cucurbits such as winter squashes are relatively dry by the time the seed is mature and ready for extraction. In this case water has to be introduced into the separator to wash the seeds free from the fruit debris.

Fig. 10.2. A field of crookneck squash ready for harvest and seed extraction.

After extraction the seeds are washed in troughs (as described for watermelon). *Cucurbita* seeds are not fermented during the cleaning process as this tends to discolour them and reduce potential germination.

When the seeds have been dried they are passed through an aspirated screen cleaner to remove pieces of dried fruit debris and any light or immature seeds.

Seed yield and 1000 grain weight

The average seed yield is approximately 500 kg/ha, but under good pollination and cultural conditions up to 1000 kg/ha can be obtained.

The 1000 grain weight is approximately 200 g depending on the cultivar.

Pathogens

The main seed-borne pathogens of *Cucurbita* spp. (pumpkin, vegetable marrow and squash) with common names of the diseases they cause are listed in Table 10.4.

Miscellaneous Cucurbits

The following crops are of special importance in some geographical areas:

Benincasa hispida (Thunb.) Cogn.: ash pumpkin, Chinese watermelon, hairy wax gourd, tallow gourd, wax or white gourd. This species is important in parts of Asia, especially the tropical areas. The mature fruit can be stored for many months, hence their culinary value. Most seed stocks are based on local selections and there are also a few named cultivars in China. Plants for seed production should be selected on vegetative characters as well as size and shape of the mature fruit.

Cucumis anguria L.: West Indian gherkin, bur gherkin. Locally important in South America, especially Brazil, also important in the West Indies. Plants for seed production are selected as early as possible according to fruit types before any significant cross-pollination can occur. Rogue, or select, when the mature fruits display the important morphological characters.

Table 10.4. The main seed-borne pathogens of *Cucurbita* species; these pathogens may also be transmitted to the crop by other vectors.

Pathogens	Common names
Cladosporium cucumerinum	Scab
Didymella bryoniae	Leaf spot
Fusarium solani f. sp. *cucurbitae*	Fusarium foot rot
Xanthomonas campestris pv. *cucurbitae*	Bacterial leaf spot
Cucumber mosaic virus	Mosaic
Musk melon virus	
Squash mosaic virus	

Lagenaria siceraria (Molina) Standl.: bottle gourd, calabash gourd, white-flowered gourd, zucca melon. The fruits from a few cultivars of this species are used in culinary preparations, but the majority are cultivated for their hard external skins or rinds which are used for a wide range of domestic utensils and ornaments. Plants for seed production should be selected on vine, foliage and fruit characters. The selections intended for culinary use should also be checked for lack of bitterness.

Luffa acutangula (L.) Roxb.: angled gourd, silky gourd, angled loofah, ridged gourd, vegetable gourd.

Luffa cylindrica (L.) Roem: dishcloth gourd, smooth loofah, sponge gourd, vegetable sponge. The young or immature fruits of these species are used as a vegetable in Asia. Some types, especially *Luffa cylindrical*, are cultivated specifically for the production of the spongy and fibrous interior of the mature fruits. Selection for seed production is based on vine and fruit characters.

Momordica charantia L.: balsam pear, bitter cucumber, bitter gourd, bitter melon. This species is locally important in India and the Far East. The fruits are pickled or used in curries and in some areas the young vine or leaves are used as a cooked green vegetable. Plants selected for seed production should have fruit characters based on the criteria of local producers and consumers. Seed crops should be isolated from domestic crops by a minimum of 100 m.

Sechium edule (Jacq.) Swartz: choyote, chayote, christophine. This is a perennial whose fruits, and sometimes tuberous roots, are used as a cooked vegetable. It is cultivated in South America, and also in other tropical and subtropical areas of the world. Replacement stocks and selections are usually made by vegetative cuttings from plants with optimum fruit and morphological characters. Each fruit bears a single seed; the seed is recalcitrant and is usually kept in the intact fruit until the next planting season.

Trichosanthes cucmerina L.: snake gourd. The fruit of this crop is very popular in Asia, Australasia, South America and locally in parts of Africa; it is cultivated in the humid tropical areas of these locations. Plants for seed production should be selected for their relatively straight and perfect fruit. This species should have a minimum isolation distance of 1000 m, with particular attention given to ensure that there is no pollen contamination from nearby smallholder and garden crops.

Further Reading

Herklots, G.A.C. (1972) *Vegetables in South-East Asia*. George Allen and Unwin Ltd, London, pp. 273–348.
Robinson, R.W. and Decker-Walters, D.S. (1997) *Cucurbits*. CAB International, Wallingford, UK.
Shinohara, S. (1984) *Vegetable Seed Production Technology of Japan, Elucidated with Respective Variety Development Histories, Particulars*, Volume 1. Tokyo. Japan, pp. 318–426.

11 *Leguminosae*

This botanical family contains a large number of genera cultivated for food and fodder. Many of the species are also grown for the production of their dried seeds, which are frequently referred to as pulses or grain legumes. In addition to the vegetable species discussed in greater detail in this chapter they include:

Phaseolus aconitifolius Jacq.	Mat bean, moth bean
Phaseolus angularis (Willd.) W Wight	Adzuki bean
Phaseolus mungo L.	Black gram
Phaseolus radiatus L.	Green gram, mung bean
Trigonella foenum-graecum L.	Fenugreek
Vigna unguiculata (L.) Walp.	Cowpea

There are also species in this botanical family which are of major significance as oilseed crops, these include:

Arachis hypogea L.	Peanut, groundnut
Glycine max (L.)	Merr. soybean

There are some cultivars of soybean which have been developed for use as a green vegetable (Chotiyarnwong *et al.*, 1996). Carter and Shanmugasundarum (1993) have discussed the value of the vegetable soybean as an underutilized vegetable crop. The vegetables dealt with in this chapter are:

Pisum sativum L. *sensu lato*	Pea, garden pea
Phaseolus vulgaris L.	Dwarf bean, French bean, green bean, snap bean
Phaseolus coccineus L.	Runner bean, scarlet runner bean
Psophocarpus tetragonobolu (L.) DC.	Goa bean, winged bean
Vici faba L.	Broad bean, field bean, faba bean
Voandzeia subterranea (L.) Thou.ex DC.	Bambara groundnut, earth bean

Peas: *Pisum sativum* L. *sensu lato*

The modern cultivars have been developed from material originally introduced in to Africa, China, Europe, India and North America from South-west Asia. This early distribution over a wide area is generally believed to be the reason for the present diversity of types. Peas are now widely cultivated in temperate regions and as a cool season crop in the tropics, especially at the higher altitudes.

Some authorities have recognized two species, *Pisum arvense* L. (field peas) and *Pisum sativum* L.(garden pea). A further subdivision of *P. sativum* has also been used, i.e. *P. sativum* var. *macrocarpon* Ser. (edible podded peas) and *P. sativum* var. *humile* Poir (early dwarf peas in which the pods are lined with a characteristic membrane). However, all the types are now generally considered to be within the species *P. sativum* that is divided into four main groups (Kelly, 1988), these are:

1. Vining peas: harvested while the seeds are still tender and are used for immediate freezing, or canning as 'garden peas'.
2. Picking peas: grown as a horticultural crop for harvesting as a fresh vegetable, also harvested while the seeds are still tender. Some cultivars in this group have tender pods which are consumed complete with immature seeds (frequently referred to as 'mange tout').
3. Combining peas: harvested for their dry seeds, (in some areas they may be hand harvested). This group has two main uses for human consumption, used either for the production of canned 'processed peas' or as dried peas. The other purpose is for animal fodder.
4. Forage peas: used for animal grazing, silage or haymaking.

Cultivar description of peas

- Season of use and suitability for specific outlets or markets, e.g. processing.
- Seed: round or wrinkled, relative size, colour (influenced by colour of testa and cotyledons which are recorded separately).
- Plant height: number of nodes to first flower.
- Stem: presence or absence of pigmentation, degree of fasciation.
- Leaf: leaflets, relative size, colour, marbling.
- Stipules: developed or vestigial.
- Flower: number per raceme, colour.
- Pod: relative colour, length, shape, form at tip (apex), degree of parchment.
- Resistant to *Fusarium oxysporum*, *Erysiphe pisi* and *Ascochyta pisi*, Seed-Borne Mosaic Virus (SbmV), Bean Yellow Mosaic Virus (BYMV) and Pea Enation Mosaic Virus (PEMV).

(A detailed test guideline (007) is obtainable from UPOV; see Appendix 2.)

Soil pH and nutrition

Peas are grown on soils with a pH 5.5–6.5. The ratio of fertilizers applied in the base dressing during seed-bed preparations depends on the nutrient status and local customs, but specialist pea seed producers tend to use a N:P:K ratio of 3:1:2 which is a higher nitrogen level than that normally used for the market crop. Work by Browning and George (1981a) has indicated that increased seed yield can be obtained with relatively high levels of N and P. In addition, seeds produced from the higher N and P mother plant regimes were found on analysis to contain higher levels of N and P. However, the seeds from the higher nitrogen regimes in the same experiments were found to be less vigorous when subjected to the conductivity test (PGRO, 1978). This indicated that nutrient regimes for seed yield differ from those required to produce high quality seed.

Peas are very susceptible to manganese deficiency especially on wet soils with relatively high organic levels. The symptoms of the deficiency are brownish hollow centres to the cotyledon seen when unripe seeds are split open. The condition is commonly called 'marsh spot'. Manganese sulfate is included in base dressings at a rate of 40–100 kg/ha where manganese deficiency is known to occur. Alternatively, manganese sulfate is applied as a foliar spray at a rate of 10 kg/ha in 200–1000 l of water but this must be applied as soon as the symptoms are diagnosed, otherwise it will be too late to improve seed quality and yield.

Another indication of low vigour in pea seeds is the incidence of 'hollow heart' and 'bleaching'. The former is characterized by a cavity of dead tissue on the adaxial surface of the cotyledons (Fig. 11.1) and the later by a yellowing of the green seeds as a result of chlorophyll bleaching (Maguire et al., 1973). It is generally accepted that both these conditions are caused by high temperatures during maturity or seed drying (Perry and Harrison, 1973). During the course of the investigations on mother plant nutrition by Browning and George (1981b) it was shown that the low nitrogen and high phosphorus regime predisposed developing seeds to hollow heart whereas the high nitrogen regimes predisposed them to bleaching.

Irrigation

Salter and Goode (1967) reviewed the literature related to the irrigation of peas and found that most workers have reported a moisture-sensitive stage from the

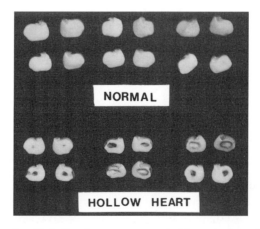

Fig. 11.1. Pea seeds with hollow heart compared with normal seeds.

start of anthesis until petal fall. This moisture-sensitive stage is very evident regardless of the amount of available water before or after flowering. Moisture deficits prior to anthesis only affect the weight of haulm produced but shortage of water during pod growth can also affect yield.

The general practice in temperate regions is, therefore, to ensure that adequate irrigation is available during and after flowering, whereas in arid areas and dry seasons in the tropics sufficient water should also be given to ensure satisfactory plant growth before flowering commences.

Sowing

Pea seeds are directly drilled into the field relatively early in the spring at the rate of 150 kg/ha in rows 40 cm apart, although sowing rate and plant density depend on the seed size and growth habit of the cultivar. Some North American seed producers use a sowing rate of up to 250 kg/ha and a much shorter distance between rows for the final multiplication stage. The higher plant densities reduce the time span of pod maturity but increase the difficulty of roguing.

Flowering

This species is a quantitative long-day plant. The earlier-cropping cultivars generally produce flowers on lower nodes than the taller and later types.

Pollination and isolation

Pea flowers are almost totally self-pollinated. This occurs in the late bud stage before the flower is completely open.

The seed purity requirements are generally higher for cultivars produced for vegetable crop production than for stock feed or other agricultural purposes.

Recommended isolation distances for peas are relatively short in many countries and aim mainly to prevent admixture during harvesting. It is important that adjacent cultivars should be at least 20 m apart with the distance increased to at least 100 m for basic seed production.

Roguing stages

1. After emergence (when plants are approximately 15 cm high): remove the taller off-types. For basic seed production particular attention is given to checking that foliage, including stipules, is typical of the cultivar.
2. Flowering: remove early flowering plants from late flowering cultivars, check flower colour and remove any plants with flower colour which is not true to type. Check flower number per node.
3. Pods formed: check that pod shape, size and colour are typical of the cultivar; remove late flowering plants; remove non- or low-yielding plants.

Harvesting

The harvesting operation normally commences when the majority of pods have become parchment-like. By this stage the maturity of individual seeds is sufficiently advanced for them not to be adversely affected by subsequent drying. Biddle and King (1977) showed that the seed quality is reduced if seeds are harvested when their moisture content is above 30–36%. However, mature seed with a moisture content of 12% or less is subject to mechanical damage.

Combining

In areas where the seeds dry sufficiently on the plants the crop is harvested by direct combining. A relatively low drum speed is used and care taken to avoid mechanical damage to the seeds.

Desiccants are used by some pea seed producers in areas where the pre-harvest drying-off of the plant material is relatively slow. They are applied when the crop starts to senesce and the lowest pods have turned pale brown and parchment-like. The usual rate of application, depending on product is 3 l in 100 l of water sprayed on at the stage when random samples of seeds have a moisture content of 40%.

Crops which have been treated with a desiccant are either combined directly when the moisture content of seeds is approximately 28%, or cut into windrows 4–5 days following application of the desiccant.

If combining is not possible due to unavailability of machines or if the production scale is too small, an alternative is to cut the crop when the earliest peas have dried to the parchment stage and the foliage is starting to dry off, characterized by a reduction in the intensity of green in the leaves and haulm. Sample pods should contain fully developed seeds which are firm, taste starchy and are readily detachable from their pods.

Windrowed crops are usually turned, although each time the windrows are turned some seed is likely to be lost by shattering.

An alternative to windrows is to cut and stack the crop on to tripods. This increases the rate of drying and reduces the time in which the seeds can deteriorate as a result of pathogen activity but the method has a relatively high labour content.

Material from windrows or tripods is threshed when the seeds' moisture content is approximately 30% using a thresher with rubber-covered tines.

Seed yield and 1000 grain weight

The average yield of a pea seed crop is approximately 2000 kg/ha.

The 1000 grain weight varies from the small to the larger seeded types. But a guideline is 330 g for the larger seeded types to 150 g for the smaller ones.

Table 11.1. The main seed-borne pathogens of *Pisum sativum*; these pathogens may also be transmitted to the crop by other vectors.

Pathogens	Common names
Ascochyta spp. including *Ascochyta pisi*	Leaf and pod spot
Cladosporium cladosporioides f. sp. *pisicola*	White mould, leaf and stem spot
Colletotrichum pisi	Anthracnose
Erisiphe pisi	Powdery mildew
Fusarium oxysporum f. sp. *pisi*	American wilt (race 1), near wilt (race 2)
Fusarium spp.	
Mycosphaerella pinodes	Foot rot, leaf spot, blight and black spot
Perenospora viciae	Downy mildew
Phoma medicaginis var. *pinodella*	Foot rot and collar rot
Pleospora herbarum	Foot rot
Rhizoctonia solani	Damping-off, stem rot
Sclerotinia sclerotiorum	Stem rot
Pseudomonas syringae pv. *phaseolicola*	
Pseudomonas syringae pv. *pisi*	Bacterial blight
Xanthomonas rubefaciens	Purple spot
Pea early browning virus	
Pea enation mosaic virus	
Pea false leaf roll virus	
Pea mild mosaic	
Pea seed-borne mosaic virus	
Ditylenchus dipsaci	Pea eelworm

Pathogens

The main seed-borne pathogens of *Pisum sativum* are listed in Table 11.1.

French Bean: *Phaseolus vulgaris* L.

Phaseolus vulgaris, which originated in North America, is cultivated throughout the temperate, tropical and subtropical areas of the world. The evolution of *Phaseolus* species under domestication has been discussed by Smartt (1976, 1990).

There are several common names for this crop and, in addition to French bean it is also referred to as common bean, snap bean, green bean, kidney bean, haricot bean and dwarf bean. The species is cultivated for either dried beans or the immature green pods. It is an important crop for processing, the immature green pods are canned or frozen; the dried beans are used in many different ways, including the preparation of canned navy or baked beans. Specific cultivars have been developed for each of these processes and 'stringless' types have also been selected which are suitable for the fresh crop and processing. The majority of cultivars are bush types but there are also climbing ('pole') types which are frequently referred to as 'climbing French bean'.

Cultivar decription of French bean

- Season and use: flowering and cropping season, suitability for specific market outlets, e.g. green, processing or dried, suitability for mechanical harvesting.
- Seed: relative length, testa colour and patterning, bars or mottle, if bicoloured, shape resistance to mechanical damage.
- Plant habit: dwarf or climbing, bush types degree of branching.
- Leaf: colour, shape, texture, size.
- Flower: colour of standard; colour of wing.
- Pod: length, including beak character, shape of transverse section (through seed).
- Seed: weight, size, shape, colour.
- Resistance to Bean Common Mosaic Virus, Bean anthracnose, Halo blight and Common blight.

(A detailed test guideline (012) is obtainable from UPOV; see Appendix 2.)

Soil pH and nutrition

Phaseolus vulgaris tolerates slightly acid soil conditions; soils with a pH 5.4–6.5 can be used successfully. The general N:P:K ratio applied during seedbed preparation is 1:2:2. Gavras (1981) investigated the effect of mother plant mineral nutrition on seed yield and quality and found that different ratios of nitrogen and phosphorus were required for relatively high seed yield and for seed quality. Browning *et al.* (1982) reported the importance of a correct balance between nitrogen and phosphorus available to the plant for the production of high vigour seeds. Figure 11.2 shows the effects of three levels, each of phosphorus and nitrogen on the yield of bean seeds as determined by the cold test. The cold test was as described by Hampton *et al.* (1995).

Irrigation

Literature reviewed by Salter and Goode (1967) indicated that water shortages during anthesis and pod development seriously affected bean yield. Work in European temperate regions showed that irrigation applied before the start of anthesis only increased vegetative growth. However, as discussed by Davis (1997), early moisture stress, at the two trifoliate leaf stage can have a detrimental effect on growth and affect flower initiation resulting in uneven crop maturity. There are many bean production areas where early and sustained irrigation is necessary to obtain a satisfactory crop.

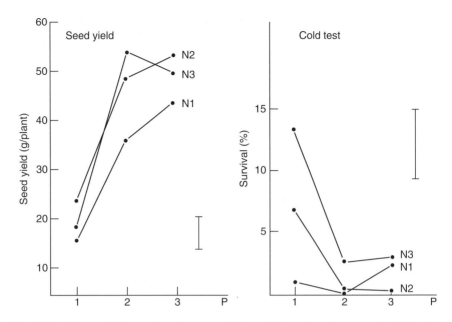

Fig. 11.2. Effects of three phosphorus and three nitrogen levels on seed yield and seed vigour in *Phaseolus vulgaris*.

Sowing

Seed of the bush types is sown in the late spring when the possibility of frost has passed and soil temperatures have risen sufficiently for satisfactory germination and plant growth. In some specialized areas in North America, notably Michigan, where length of season allows double cropping, seeds of early-maturing cultivars are drilled into the stubble of a previous barley crop or immediately after the stubble is ploughed in. When this technique is adopted seeds of early-maturing cultivars are drilled in early July.

The sowing rate depends on the relative seed size of the cultivar but is approximately 100 kg/ha for the bush types and 50 kg/ha for the climbing types. Bush cultivars are sown in rows 45–90 cm apart, according to cultivar and stage of multiplication. The climbing cultivars are sown in rows 90–120 cm apart. In some tropical areas cane frames or supports are erected, although generally they are supported by strings.

Flowering and pollination

Most cultivars of each type (i.e. bush and climbing) are day-neutral although there are some short-day cultivars in each group. The physiology of *Phaseolus* beans, including induction of flowering, pollination and pod growth has been reviewed by Davis (1997).

The flowers are self-compatible and are predominantly self-pollinated although some cross-pollination occurs. The degree of cross-pollination probably increases in the tropics where insect activity (including thrips) is greater (Drijfhout, 1981).

Isolation

Many seed authorities stipulate a minimum distance of 3 m between a seed crop and any other crop of *Phaseolus vulgaris*. However, 50 m for seed crops in the final multiplication stage and 150 m for a crop intended for basic seed are preferable isolation distances.

Roguing stages

1. Before flowering: check plant habit, vigour and height according to type; check foliage, leaf shape and colour.
2. At onset of flowering: check plant vigour and flower colour; remove plants showing symptoms of seed-borne pathogens.
3. Seed set and first pods formed: check pod characters, including shape and colour; remove late flowering off-types and plants showing symptoms of seed-borne pathogens.

Basic seed production

A scheme for maintaining breeders' stocks of beans has been described by Drijfhout (1981). For basic seed single plant selections are made from a relatively large plant population grown from breeders' stock seed. The selections are based on five successive inspections commencing immediately after seedling emergence. Particular care is taken to retain only seedlings with appropriate cotyledon colour. The young plants are then inspected a second time to remove plants which show inappropriate leaf characters. The third selection is made just prior to flowering in order to confirm the general plant form, height and earliness.

When flowering commences, the plants are examined a fourth time to check flower colour, and off-types are again removed. The fifth selection is made when the pods are developing and characters such as pod shape, length and colour can be seen, and other characters such as stringlessness can be detected. Plants showing symptoms of seed-borne pathogens are discarded at each stage of selection.

The remaining plants are harvested singly and low yielding plants (as determined by seed yield) are rejected. The remaining seeds from selected plants are sown the following season as single plant selections and undesirable lines are rejected. The remainder are bulked for further multiplication and used as basic stock.

Harvesting

The dwarf or bush types are generally considered to be ready for a once-over harvest when the earliest pods are dry and parchment-like, and the remainder of the pods have turned yellow. The seeds' moisture content at the time of harvesting should be between 20 and 25%; seed maturity is confirmed by opening sample pods, wherein the seeds should be fully developed with a mealy texture. Under good growing conditions the flowers tend to set until relatively late in the season. This leads to a loss from 'shattering' of the earliest mature seeds if harvesting is delayed. Smith (1955) examined the effects of stage at harvesting on bean seed yield and quality and found that there was a reduction equivalent to 358 kg/ha in harvested seed when the crop was cut before the earliest pods were fully mature. Figure 11.3 shows dwarf bean plants with mature seed pods prior to harvesting.

The plants are cut and placed in windrows for further drying before either combining or threshing, or they are combined direct from the standing crop. The entire operation is planned to ensure both the minimum loss from shattering and the least possible mechanical damage to the seeds which are especially susceptible to cotyledon cracking.

Fig. 11.3. Dwarf bean plants with ripe pods prior to cutting and harvesting.

The climbing cultivars mature over a longer period than the bush types so they are harvested by hand on three or more successive occasions as the older pods mature, or the plants are pulled out and dried off in windrows.

Threshing

Small quantities of seed are threshed by hand to avoid subsequent loss due to mechanical damage; this is especially important with small seed lots of basic or stock seed. Large seed lots are threshed with a drum speed of 250–350 rpm at a concave clearing of *c.*12–20 cm. The seeds' moisture content should not be too low or excessive mechanical damage occurs during machine threshing.

Wilson and McDonald (1992) have evaluated six systems for the threshing of *Phaseolus vulgaris* and found that the highest quality seed was produced by open flail threshing or hand shelling. They concluded that, where feasible, manual threshing methods are superior to mechanical methods for small seed lots. Manual threshing would be expected to minimize cotyledon cracking.

Seed yield and 1000 grain weight

The seed yield is *c.*1500 kg/ha although under ideal production and harvesting conditions yields are *c.*2000 kg/ha.

The 1000 grain weight for *Phaseolus vulgaris* is *c.*250 g, although this can be up to 600 g in the smaller seeded cultivars.

Pathogens

The main seed-borne pathogens of *Phaseolus species* are listed in Table 11.2.

Runner Bean: *Phaseolus coccineus* L.

Origin and types

The runner bean, also commonly called 'scarlet runner bean', originates from Central America but is now widely distributed in the temperate regions. It is especially popular in Europe, where it is mainly cultivated as a garden vegetable. It is a perennial with a tuberous root stock but because of its sensitivity to frost it is usually cultivated as an annual.

The main product is its edible partly developed green pods, but in some areas the mature seeds are also used for culinary purposes.

Most cultivars are of the twining form and are normally grown on supports. There are a few non-climbing types, often referred to as 'dwarf' cultivars which are mainly used for large-scale commercial crop production.

Table 11.2. The main seed-borne pathogens of *Phaseolus* species; these pathogens may also be transmitted to the crop by other vectors.

Pathogens	Common names
Alternaria alternata	
Ascochyta boltshauseri,	
Ascochyta phaseolorum and	
other *Ascochyta* species	Ascochyta leaf spots
Aspergillus sp.	'Baldheads' and 'snake-heads' of seedlings
Cercospora canescens	
Colletotrichum lindemuthianum	Anthracnose
Fusarium oxysporum f.sp. *phaseoli*	Yellows, wilt
Fusarium solani f. sp. *phaseoli*	Root rot
Phaeoisariopsis griseola	Angular leaf spot
Pleospora herbarum	Red nose, leaf spot
Rhizoctonia solani	Damping-off, stem canker
Sclerotinia sclerotiorum	Sclerotinia wilt, stem rot, watery soft rot, white mould
Pseudomonas syringae	Bacterial brown spot
Pseudomonas syringae pv. *phaseolicola*	Halo blight, grease spot
Xanthomonas campestris pv. *phaseoli*	Common bacterial blight, fuscous blight bean common mosaic virus
Cherry leaf roll virus	
Cucumber mosaic virus	
Runner bean mosaic virus	

Cultivar description of runner bean

- Cultivars are grouped according to their flower colour.
- Flower colour; white, pink, red or red standard and white wing petals.
- Use: fresh market, processing, freezing, private garden, including suitability for exhibition.
- Plant: growth habit; bush, dwarf or climbing.
- Pod: length short, medium or long.

(A detailed test guideline (009) is obtainable from UPOV; see Appendix 2.)

Agronomy

The soil and nutrient requirements are the same as described above for *P. vulgaris*. Seed is sown at a rate of up to 100 kg/ha when grown on supports, in rows 100 cm apart. In large-scale production of the climbing types the final multiplication is done without plant supports, but for basic seed production the plants are grown on individual strings to facilitate roguing. The dwarf types are grown at closer row spacing similar to *P. vulgaris* but with approximately 50% increase in sowing rate because of the larger seeds of *P. coccineus*.

Irrigation

The main work on the response to different soil moisture conditions at different stages of growth has been done by Blackwall (1961, 1962, 1963). Some of Blackwall's investigations included a study of the interaction of moisture status with organic manure treatments. The application of farmyard manure during soil preparations, followed by irrigation from the visible early green flower bud stage increased the early yield of green beans by 57%. Withholding irrigation until the flowers showed their colour increased yields by only 16%. These increases were obtained in comparisons with runner beans grown in plots with no additional farmyard manure or irrigation.

Flowering

The runner bean is a long-day plant but does not set pods when temperatures are higher than 25 °C. It is also similarly sensitive to night temperatures below 20 °C which tend to cause abscission of both flowers and young pods (Smartt, 1976).

Pollination

P. coccineus is self-fertile but some insect activity is required to 'trip' the stigma (i.e. rupture the stigmatic surface slightly). This is usually achieved by honeybees (*Apis mellifera*) or bumble bees (*Bombus* spp.). There is up to 40% cross-pollination, especially early in the flowering season. Some species of bumble bees (*B. lucorum* and *B. terrestris*) are short-tongued and obtain nectar by making holes near the base of the flowers (Free and Racey, 1968); they are, therefore, not efficient cross-pollinating agents.

Isolation

A minimum isolation distance of 100 m is stipulated by most authorities, although this should be increased for stock seed production. Some seed regulatory authorities stipulate a greater isolation distance between cultivars with different flower colours.

Roguing

The roguing stages and associated characters to observe are similar to those described for *P. vulgaris*. Particular attention should be given to the early removal of off-types before they have contributed to the pollination of other plants in the seed crop.

Harvesting and threshing

The harvesting methods are similar to those described for *P. vulgaris* although a greater degree of hand harvesting is done from plants grown on supports.

Runner beans are threshed with a drum speed of 250–350 rpm and a wide concave opening to minimize mechanical damage.

Seed yield and 1000 grain weight

The average seed yield is *c.*1000 kg/ha.

The 1000 grain weight is *c.*1 kg, although the weight of some smaller seeded cultivars is less.

Pathogens

The main seed-borne pathogens of *Phaseolus* species are listed in Table 11.2.

Broad Bean: *Vicia faba* L.

This species is also commonly referred to as field, horse, tick and Windsor bean, although all but the latter common names usually refer to cultivars not used as vegetables for human consumption.

Vicia faba probably originated in west or central Asia (Pilbeam and Hebblethwaite, 1994). It is widely cultivated in temperate areas and in the tropics, subtropics and arid areas as a cool season crop. This species is cultivated as stock feed (as mature seed, hay or silage) and as a vegetable. The vegetable crop is either produced for the slightly immature seeds which are cooked fresh or processed. In some arid and tropical areas the dried mature seeds are stored for use in the winter and other seasons when fresh vegetables are in short supply. The origins, yield variability, breeding and prospects for the species have been discussed by Pilbeam and Hebblethwaite (1994).

Earlier classifications of the range of cultivars used by agriculture and horticulture were based on their agronomic uses; however, as a result of the large degree of overlap between the different types this is no longer of value.

Cultivar description of broad bean

- Season and use: suitability for early sowing, winter hardiness and maturity period.
- Seed: weight of 1000 seeds, relative size; testa and hilum colour; tannin absent or present.
- Plant: height; number of tillers.
- Leaf: stipules, pigmentation, melanin spots.

- Flower: wing petals, pigmentation on standard petal; presence or absence of spot on the reverse.
- Pod: length, breadth, approximate number of seeds per pod.

(A detailed test guideline (206) is obtainable from UPOV; see Appendix 2.)

Soil pH and nutrition

Soils for broad beans should have a pH of 6.5 with adequate available calcium, otherwise a suitable liming material must be applied during soil preparations.

The nutrient ratio applied during the final stages of seedbed preparations is an N:P:K ratio of 1:1:1. It is particularly important not to apply excessive nitrogen, especially for autumn sown crops.

Sowing

The long-pod group of cultivars which have some frost resistance are sown in the late autumn but in areas with severe winters, sowings are made in early spring as with the other types of broad bean.

Seeds are sown at the rate of 150 kg/ha in double rows 25 cm apart with 70 cm between the rows.

Irrigation

Broad beans are very responsive to irrigation carried out during anthesis. Brouwer (1949, 1959) found that dry soil at this stage had the greatest adverse effect on yield, while irrigation before the onset of anthesis had relatively little advantage even under dry conditions. However, Jones (1963) found that seed yield was to some extent dependent on the plants having a relatively high growth rate before flower development. It is, therefore, considered preferable to ensure that the plants receive sufficient water during early development and also from the start of flowering.

Flowering

The majority of broad bean cultivars require long days for the acceleration of flower initiation, although this is not so with the earliest-flowering cultivars (Whyte, 1960). There is also evidence that vernalization at temperatures below 14 °C will also accelerate flowering.

Pollination

Broad beans are self-compatible, but both self- and cross-pollination occur. Insect activity is responsible for up to 30% crossing (Watts, 1980). However, as with *Phaseolus coccineus,* some bees obtain nectar by making a hole at the base of the flowers as the season advances, thus reducing the incidence of cross-pollination.

Reisch (1952) found that high rainfall during flowering prevented pollination by insects and thereby was responsible for a considerable reduction in yield.

Isolation

Most authorities stipulate a minimum isolation distance of 300 m between broad bean crops. However, because of the relatively high incidence of cross-pollination, isolation distances should be at least 1000 m, especially for stock seed production. The greater isolation distances are especially important for cultivars used to produce horticultural crops as a higher degree of cultivar purity is desirable.

Roguing stages

1. Before the start of anthesis, check: (i) general plant habit vigour, height, tiller number; (ii) stipules: presence or absence of melanin spot; (iii) incidence of seed-borne pathogens, remove infected plants.
2. At start of anthesis, check: (i) general plant morphology; (ii) flower colour, including wing petals and standard; (iii) incidence of seed-borne pathogens, remove infected plants.
3. When pods set, check: (i) colour, shape and relative length of pods; (ii) pose of pods.

Harvesting and processing

Signs indicating that a broad bean seed crop is ready for harvest include the pods becoming relatively dry with a loss of sponginess; this is preceded by a general blackening of the pods. The optimum seed moisture content at harvest is 16–20%.

The crop is either cut by hand and tied in bundles or mown by machine. Bundling and stooking is necessary in northern Europe, although not in the drier seed production areas such as the Middle East and north Africa.

The material is threshed when dry at a drum speed of 250 rpm. Care must be taken to adjust the concave setting according to the seed size of the cultivar to minimize mechanical damage to the seeds.

Seed yield and 1000 grain weight

The seed yield is *c*.1500 to 2000 kg/ha. The 1000 grain weight is *c*.800–1200 g for the smaller seeded cultivars and *c*.2000 g for the larger-seeded cultivars.

Table 11.3. The main seed-borne pathogens of *Vicia faba*; these pathogens may also be transmitted to the crop by other vectors.

Pathogens	Common names
Ascochyta fabae	Leaf and pod spot
Botrytis fabae	Chocolate spot
Colletotrichum lindemuthianum	Anthracnose
Fusarium spp.	Foot rots, wilts
Pleospora herbarum	Net blotch
Uromyces viciae-fabae	Rust
Bean yellow mosaic virus	
Broad bean mild mosaic virus	
Broad bean stain virus	
Echtes ackerbohnenmosaik virus	
Ditylenchus dipsaci	Stem eelworm

Pathogens

The main seed-borne pathogens of *Vicia faba* are listed in Table 11.3.

Bambara Groundnut, Earth Nut: *Voandzeia subterranea* *L. Thou. ex DC.* (syn. Ground Bean, Madagascar Groundnut: *Vigna subterranean* (L.) Verdc.)

Origin and uses

This species is cultivated in west Africa, parts of South-east Asia, especially the Philippines and Indonesia, and limited areas of South America particularly in Brazil and Surinam.

It is predominantly grown by subsistence farmers, especially on poor soils in areas with high temperatures and low rainfall. Linnemann and Azam-Ali (1993) have regarded it as an underutilized crop with important potential for improving food security.

Bambara is cultivated for its edible seeds, which are usually consumed before they ripen, although the ripened seeds are also ground to flour for culinary preparations. The seeds are produced under the ground as a result of the recurving inflorescences pushing into the soil.

Morphological types

There are no named cultivars cited in the literature, although local selections are maintained by farmers in some areas. The types which are generally considered more suited to cultivation are those which have less of a trailing habit and form plants which are more clumped. Hepper (1970) considers that there are two botanical varieties in cultivation, var. *spontanea* which includes wild forms and var. *subterranea* which includes the cultivated forms.

According to Doku (1969) selections are made for short internodes, short or tall plant types, plant habit including spread, flower colour (yellow, green or orange), number of seeds per pod, pod size and colour. Descriptors have been published by IBPGR (1987).

Soil pH and nutrition

The crop is suited to light soils that facilitate harvesting. Seedbeds should be prepared to form a fine tilth. The optimum pH is in the range of 5.5–6.5 although the crop requires adequate available calcium for satisfactory development of the nuts. The optimum N:P:K fertilizer ratio is 1:2:2 applied during site preparation. The majority of subsistence farmers do not have bulky organic manures available, but the crop does respond to them. If used, due attention should be made to reduce the nitrogen in base dressings otherwise the crop will be very vegetative. Johnson (1968) reported that high nitrogen regimes should be avoided.

Sowing and crop establishment

The unshelled seed can be sown, although it is generally considered preferable to shell before sowing. Shelled seed is sown approximately 7 cm deep in flat-topped ridges at the rate of 40–50 kg/ha, at stations 50–60 cm apart with 50 cm between the ridges. It is especially important to use a ridged system where waterlogging is likely to occur otherwise a flat seedbed may be used. The plants are earthed up during anthesis which commences approximately 30–50 days from sowing.

Harvesting

The crop should be lifted from the soil as soon it is ready, otherwise the seeds may commence germination, especially in areas with a following rainy season. The maturity of the subterranean pods should be determined by spot examinations from the time that the aerial foliage commences to turn yellow. Under dry conditions the leaves will also wilt and wither as the seeds reach maturity.

The mature seeds are either ploughed out or hand dug, depending on the scale of operation. The lifted plants with pods are left in windrows on the soil surface to further dry for up to 3–5 days.

The seed pods are either hand threshed or the seed is extracted with a groundnut thresher. The shelled seed should have its moisture content reduced to 13%, although for vapour-proof storage it is reduced to 11%.

Yield and 1000 grain weight

The average seed yield is 600 kg/ha.
The 1000 grain weight is approximately 670 g.

Seed-borne pathogen

The only seed-borne pathogen listed by Richardson (1990) is an unidentified virus which causes a mosaic.

Winged Bean, Asparagus Bean, Goa Bean: *Psophocarpus tetragonobolus* (L.)DC.

This species is an established smallholder and home-garden crop in the more humid areas of South-east Asia. It has been identified as potentially suitable for wider adoption in the tropics and some temperate areas, especially for subsistence farmers.

The plant provides a range of culinary uses; the immature pods are used as a green vegetable when about half-developed, the leaves can be used as a leafy green vegetable, the tuberous roots may be eaten and the mature seeds are cooked following a specified preparation.

The winged bean plant is a climbing perennial, but is usually treated as an annual.

Origin and types

There is some speculation regarding the origin of this species but it has long been established as a cultivated crop in Papua New Guinea, Indonesia and Madagascar.

Named cultivars are only of local significance, but the following characteristics can be used in the description of local cultivars and germplasm. IBPGR (1982) has published a descriptor for this species. The following can be used as an outline for descriptions of selections and also cultivars developed and maintained:

- Seed: colour when mature, relative size, shape, surface texture and hilum.
- Vegetative characters: stem, pigmentation, vigour, amount of growth by anthesis.
- Leaf: size and shape of leaflets.
- Inflorescence: days to start of anthesis, flower colour.
- Pod: relative length, width, shape of cross-section, colour, surface texture.
- Time to maturity.

Agronomy

There is little available information on the crop's nutrient requirements, but it responds to bulky organic manures used in soil preparation. A base nutrient dressing incorporated in the seedbed of N:P:K of 1:6:4 would take into account its response to a low nitrogen-to-phosphorus ratio, although a higher level of nitrogen will be indicated when there is significant leaching.

Seed is sown in rows 1 m apart with a final distance of 50–60 cm between plants within the rows. Approximately 7 kg of seed will sow 1 ha.

The plants are provided with vertical supports to a height of approximately 2 m.

Isolation

The minimum recommended isolation distance is 500 m for this species which is partly self-pollinated and partly cross-pollinated. However, some seed regulatory authorities may stipulate greater isolation distances.

Roguing

The crop should be rogued either at the start of anthesis, or as soon after the start as possible. Individual plants should be checked for trueness to type. A further roguing should be done when the first pods are formed and display their morphological characters.

Harvesting and threshing

The seed crop is usually hand harvested. The seeds in an individual pod are sufficiently mature for harvesting when their pod has become fibrous and started to turn brown. However, by the time that pods have turned completely brown there is a high risk of shattering. The duration from flowering to this stage depends on the cultivar, but is generally 6–7 weeks. The harvested pods are dried further before seeds are extracted by hand. The extracted seed is dried down to 13.5% moisture content for storage, or 11.5% for vapour-proof storage.

Seed yield and 1000 grain weight

Seed yields are between 800 kg/ha to 1.1 t/ha.
The 1000 grain weight is *c.*500 g.

Seed-borne pathogen

The only seed-borne pathogen listed by Richardson (1990) is Psophocarpus Ringspot Virus which is stated to be related to, but serologically distinct from, Cucumber Mosaic Virus.

Further Reading

Legumes

Powell, A.A. and Matthews, S. (1981) The significance of seed coat damage in the production of high quality legume seeds. *Actae Horticulturae* 111, 227–274.

Peas

Matthews, S. (1973) The effect of time of harvesting on the viability and pre-emergence mortality in soil of pea (*Pisum sativum*) seeds. *Annals of Applied Biology* 73, 1–9.

Dwarf bean

Gavras, M. and George, R.A.T. (1981) A review of the effects of mineral nutrition on seed yield and quality in *Phaseolus vulgaris. Actae Horticulturae* 111, 191–194.

12 *Solanaceae*

The genera in this botanical family provide some of the most important vegetables in the world, including:

Lycopersicon lycopersicon (L.) Karsten ex Farw. syn. *Lycopersicon esculentum* Mill.	Tomato
Solanum tuberosum L.	Potato
Solanum melongena L.	Aubergine, eggplant
Capsicum annuum L.	Sweet pepper
Capsicum frutescens L.	Chilli pepper

There are several other vegetable species in this family including *Physalis peruviana* L. (Cape gooseberry), *Cyphomandra betacea* (Cav.) Sendt. (tree tomato), *Solanum quitoense* Lam. (naranjillo) and *Solanum muricatum* Ait. (melon pear or pepino). Other economically important species in this botanical family that are not vegetables include *Nicotiana tabacum* L. (tobacco) and some other poisonous akaloid-bearing drug plants including *Atropa belladona* L. and *Datura stramonium* L. Although these last three species are clearly not vegetables, it is as well to realize that as related crops they are important hosts to the viruses which infect a wide range of genera in *Solanaceae* and should be taken into account when considering isolation requirements.

There are a few *Solanum* species collectively known as 'African eggplants' which are mainly used as leaf vegetables, especially locally in parts of Africa. These include *Solanum macrocarpon* L. (Gboma aubergine) which is important in West Africa, and *Solanum aethiopicum* L. (scarlet aubergine) used in Central Africa. Grubben (1977) discusses the role of these and better known *Solanum* species. As far as is known to the author there is no commercial seed production of these two species. IBPGR (1988) has published descriptors for aubergine, including the wild forms.

Tomato: *Lycopersicon lycopersicon* (L.) Karsten ex Farw. syn. *Lycopersicon esculentum* Mill.

Origins and types

The genus *Lycopersicon* originated in the narrow west coast area of South America extending from Ecuador to Chile, between the Andes and the sea, except *Lycopersicon cheesmanii* Riley, which is found in the Galapagos Islands. Evidence indicates that material of *L. lycopersicon* was introduced into Mexico where it underwent further domestication and selection which assisted in the more recognizable characters of present-day tomatoes before a wider distribution took place (Jenkins, 1948).

There are two main types of tomato plants, the determinate (or bush) and indeterminate, trained as a single stem ('vine'), by frequent removal of the side shoots as they emerge from the leaf axils.

The determinate or 'bush' tomato cultivars are widely grown in the world and are used for the majority of the field crops grown for the fresh market or processing, such as canning, purée, soup, juice or ketchup. Cultivars suitable for modern mechanical harvesting systems have been developed by plant breeders. These processing cultivars, which are usually mechanically harvested, exhibit many important characters, such as a large number or fruits set at any one time which subsequently ripen over a short period, longevity of ripe fruit on the plant, the so-called jointless character (which enables the fruit to detach from the calyx where it joins the fruit rather than the abscission layer or 'knuckle'), and a high dry matter content of the fruit.

Similarly, the modern indeterminate cultivars have been largely produced by plant breeders for the protected cropping industry where there is a high capital investment per unit area, they are also used in some field production systems where high quality fresh fruits are required, usually for salad or 'slicing', and where there is adequate hand labour for training the plants and picking the fruit over a prolonged marketing period. The indeterminate types are also popular among private or home gardeners, especially in Europe and North America, although there has been trend for seed houses to make improved determinate small-fruited cultivars more available to home gardeners.

Many other characters have been included in breeding programmes of both the determinate and indeterminate types of tomato cultivars, probably the most important of which are resistance to specific pathogens and pests with improvement in fruit qualities, such as shape, size, and even-ripening; this last character reduces the incidence of green shoulders of the ripening fruit. Plant breeders have also successfully combined desirable characters in F1 hybrids produced for specific environments and markets for extensive field or intensive greenhouse production.

Cultivar description of tomato

- There are open-pollinated and F1 hybrid cultivars.
- Plant: growth type; determinate or indeterminate.

- Foliage: attitude: semi-erect, horizontal or drooping.
- Leaf:
  - Length short, medium or long.
  - Width narrow, medium or broad.
  - Division of blades; pinnate or bipinnate.
- Peduncle: abscission layer, absent or present.
- Fruit: size very small, small, medium, large or very large.
  - Shape in longitudinal section: flattened, slightly flattened, round, rectangular, cylindrical, heart-shape, obvoid, ovoid, pear-shape or strongly pear-shape (the basic tomato fruit shapes are shown in Fig. 12.1).
  - Ribbing at stem end absent or very weak, weak, medium, strong or very strong.
  - Predominant number of locules 2, 2 and 3, 3 and 4, more than 4.
  - Green shoulder before maturity absent or present.
  - Colour at maturity yellow, orange, pink or red.
- Time of maturity: early, medium or late.
- Resistance to specific pests and/or pathogens, including: *Meloidogyne incognita, Verticillium* sp., *Fusarium oxysporum* f. sp. *lycopersici, Fusarium oxysporum* f. sp. *radicis lycopersici, Cladosporium fulvum, Phytophthora infestans, Pyrenochaeta lycopersici, Stemphylium* spp., *Pseudomonas tomato, Pseudomonas solanacearum*, Tobacco Mosaic Virus and Tomato *Yellow Leaf Curl Virus*.

(A detailed test guideline (044) is obtainable from UPOV; see Appendix 2.)

Agronomy

The plants are raised in nursery beds and planted bare-rooted into the field, or seedlings are raised and potted individually into suitable containers before they are planted into their final quarters.

There is little direct drilling or sowing into the field anywhere in the world. The practice of raising young plants as a separate operation enables the grower to achieve a high degree of plant uniformity and also provides a check against early pests and diseases while the plants are still in the nursery. The use of nursery beds also allows an early start to the crop under plastic or glass protection before planting out.

The final stand and density of plants in the field depends on whether the cultivar is a trained or bush type. The traditional spacing for staked or 'trellis' plants is 30–60 cm within the rows and 90–120 cm between the rows. But the final decision

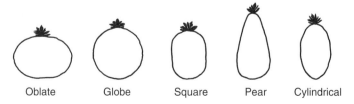

| Oblate | Globe | Square | Pear | Cylindrical |

Fig. 12.1. The basic fruit shapes of tomato.

on plant population will depend on the cultivar, irrigation system and whether or not a support system is used on which to train the plants. In greenhouse production the plant population is approximately 3000 plants/ha.

There are approximately 400 seeds/g for most cultivars, although those used by the greenhouse industry tend to be larger seeded, with approximately 300 seeds/g.

Sowing rate

For transplants, 60 g of seeds are drilled per 250 m² of seedbed which will provide sufficient plants for 1 ha. When direct drilled, the sowing rate is between 500 and 1000 g/ha.

Soil pH and nutrition

The soil pH should be maintained at 6.5; appropriate quantities of liming material must be added during preparation if the pH is lower than this. The disorder commonly called 'blossom-end rot' or 'black spot' is the result of calcium deficiency, although even when there is sufficient calcium available from the soil or substrate, the condition can arise when the soil moisture level has been allowed to become low for about 2 or more days. When several fruit or trusses are affected, seed yield will be significantly affected. The symptoms of blossom-end rot are shown in Fig. 12.2.

The nutrients applied during field preparation are calculated according to the existing status of the soil. The following amounts (all kg/ha) are applied on soils of relatively low fertility: N 75–100, P_2O_5 150–200 and K_2O 150–200.

Fig. 12.2. Symptoms of blossom-end rot on field grown tomato fruit.

A relatively high level of available phosphorus is important for a satisfactory tomato seed yield and quality when plants are grown in glasshouses (George et al., 1980). But while a high nitrogen level has been shown to produce high quality seeds, under field conditions, the nitrogen status should be about half the potassium status in order to maintain a satisfactory balance between inflorescences and vegetative growth.

Flowering

The tomato is normally grown commercially as an annual crop, although under continued ideal conditions it tends to be a relatively short-lived perennial.

The plant cannot survive temperatures at or below freezing and even prolonged periods of less than $10\,°C$ will significantly reduce production. It can survive relatively high temperatures and continues to grow very successfully up to about $25\,°C$; at higher temperatures than this there can be an adverse effect on growth rate and fruit yield even though the plant appears to be in a satisfactory vegetative condition, mainly because of infertile pollen.

The flower initiation is not dependent on photoperiod, although there is a tendency for flower truss initials to abort in the relatively low winter light levels experienced, for example, in the greenhouses of northern Europe. The plant responds well to high light levels.

The physiology of this crop species, including flowering, development to anthesis, subsequent fruit growth and development has been discussed by Kinet and Peet (1997).

Pollination

The tomato is generally considered to be a self-pollinating crop. Its flower structure favours complete self-pollination without the need for insects or other natural pollinating agents; most cultivars, especially those developed for greenhouse crop production, have continued to maintain this trait. However, some cross-pollination can occur for one of several reasons. First, there is always the possibility of an appropriate insect introducing foreign pollen from outside the crop, which is increased when flowers have been emasculated in preparation for F1 hybrid seed production. Another factor is the presence of insects capable of cross-pollination. These insects are much more likely to be encountered when tomatoes are produced in or near their original habitat of South America (Richardson and Alvarez, 1957). The possibility of insects with similar capabilities being encountered in other seed production areas cannot be ruled out. Third, despite the trend of plant breeders and seed producers to select towards complete self-pollination, there are some cultivars whose flowers have long styles which tend to favour cross-pollination. This factor and its effect on natural cross-pollination in tomatoes were first discussed by Lesley (1924). The majority of modern tomato cultivars have short-styled flowers.

There is a noticeable reduction in fertilization when ambient temperatures reach about 42 °C. This occurs in tropical field crops when high temperatures persist for long periods. The poor setting resulting from high temperatures is not only attributed to a decrease in pollen fertility, but also to physiological factors affecting fertilization. Under greenhouse conditions it is considered necessary to agitate the plants daily to assist the transfer of pollen from the anthers to the stigmas. A useful confirmation that pollination is taking place at a satisfactory rate is when only two successive flowers can be seen open on a truss at any one time. Preceding flowers on the truss would have been pollinated and succeeding flowers will still be in bud.

The period from fertilization to ripe fruit will depend on the cultivar and environment. This period can be 40–60 days but the usual time expected in most tomato production areas is about 45 days.

Isolation

The minimum isolation distance between different cultivars of tomatoes for seed production is relatively short. This is because of the crop's high level of self-pollination. Most countries specify a distance of between 30 and 200 m, the main consideration being that the distance should be sufficient to avoid admixture at harvesting time.

The distances between two parental lines grown for the production of a specific F1 hybrid need be no more than 2 m, but may be less if there is a physical barrier such as in greenhouses.

Previous cropping

The area should not have grown tomatoes in the previous 2 years, but if soil sterilization using heat or a suitable recommended soil sterilant is carried out after each crop, then successive crops can be grown in the same area. Regular partial soil sterilization or soil fumigation is practised in greenhouses where rotation is less practicable and the overall investment is much higher than field production. The seed regulatory authorities in some countries insist that no solanaceous crop should have been grown for at least the previous 3 years. This is more from the point of view of preventing pest and disease build-up than reducing the possibility of seedlings from previous crops as there are important soil-borne pests and pathogens of species in the family Solanaceae, including nematodes (eelworms) and some of the wilt diseases.

Roguing stages

1. Before flowering: note the desirable characters; growth habit and leaf morphology typical of the cultivar, observe if specific pathogens are present.

2. Early flowering and first fruit set (but before significant fruit growth): general plant morphology and characters as defined for stage 1. Check that immature fruit characters are according to the cultivar description, e.g. greenback or greenback-free.

3. First ripe fruit: yield potential, fruit morphology, colour when ripe, relative size.

Production of F1 hybrid seed

The majority of hybrid tomato seed used in the world is produced under field conditions. However, some hybrid seed of indeterminate cultivars is produced in greenhouses for the specialized protected cropping industry.

The production of hybrid tomato seed involves the maintenance and growing of two separate parental lines of plants, i.e. the male and female parents. The male is used as a source of pollen and the female receives the pollen and subsequently produces fruit containing the hybrid seed. Thus, a separate batch of plants of each parent is grown. The seed producer grows the two parents and does the hybridization in accordance with the breeder's instructions.

It is normal practice to sow the male parent up to 3 weeks earlier than the female plants to ensure an adequate supply of pollen from the start of anthesis on the female plants. Local observations of the time from sowing until start of flowering on each parent line will provide evidence for sowing dates in future years.

The ratio of male-to-female line plants will depend on the flowering habits of the individual lines, but a male-to-female ratio of approximately 1:5 is a general guideline.

Before hybridization commences all plants in both lines to be used in the programme should be checked for trueness-to-type; any off-types, undesirable or suspect plants should be removed.

The female parent plants of indeterminate types are normally supported by an appropriate system according to local conditions and custom. Potential total plant and fruit weight must be taken into account in the design of the support system. It is usually considered unnecessary to support the male parent, except when indeterminate male types are grown in structures or when a supply of pollen is likely to be needed over a long period.

The flowers on the female line must first be emasculated during their late bud stage in preparation for cross-pollination. This involves removal of their anthers, and is done as a separate operation prior to applying pollen to their stigmas. The anthers which are fused in a cone are cleanly removed by a pull using forceps. This operation requires practice and new workers should be given the opportunity to develop and practise their skill before the hybridizing season starts. It may be necessary to dissect off a single anther from the fused anther cone before being able to pull off the remainder in a single pull. It is extremely important not to damage the stigma at any time during the operation otherwise it will not be receptive to pollen and fertilization will not occur.

The timing of emasculation in relation to the opening of individual flowers is related to the rate at which flowers develop. Experience under local climatic and environmental conditions indicates how long in advance of 'normal' opening the flowers should be emasculated. For example in Californian field production, flowers are emasculated early in the morning (from about 6.30 a.m.) on the same day as pollination, whereas in northern European greenhouse hybrid seed production, flowers are emasculated up to 2 days in advance of pollination.

One of the most useful assets in hybrid tomato seed production would be to have a female parent which is male sterile, but unless a seed-producing organization has suitable parent material, hybrid tomato seed is produced by hand emasculation.

Pollen is collected from the male line either by detaching individual flowers or mechanical means such as a small vibrator. The instrument is used to agitate individual open flowers which release pollen on to a small receptive plate, disc or tube attached to the instrument. The collected pollen is gently applied with a fine artist's brush to each emasculated (or male sterile) flower. In some Asian countries, specializing in the large-scale field production of hybrid tomato seed, workers transfer the pollen from the collection receptacle to the stigmatic surface by finger. Each pollinated flower is marked with a coloured tag and some seed producers remove part of the calyx from flowers immediately after pollination; this later provides a clear confirmation of fruits containing hybrid seed at harvest time.

Under field conditions in arid areas the whole operation usually takes place three times a week on alternate working days. In temperate greenhouse production the plants may be looked over daily to emasculate flower buds at the correct stage. Any unhybridized flowers are removed or remain on the plant unmarked F1 hybrid tomato seed production is described in detail by Opeña and Chen (1993).

Every care must be taken to clearly indicate and separately label plants of either parent if they are to be maintained by natural selfing.

Harvesting and seed extraction

Seeds are extracted from ripe fruits which have either been hand-picked into containers, or collected by a single mechanized harvester that removes all the fruit from the crop in a single operation. The method of picking depends on the scale of operation and level of technology available in each seed production area.

Fruit resulting from hybridization is always hand-picked from the female lines to ensure that only those from cross-pollination are included. They are identified from either their label or calyx marking. This check, or confirmation, at the time of harvesting is extremely important when hybrid seed is being produced in order to exclude fruit which contains seed produced by natural selfing.

Seed extraction of small quantities of fruit

The ripe fruits are cut equatorially and the seeds with the gelatinous material surrounding them are squeezed or spooned out into containers. During this process the main fruit walls, pulp, skin and other debris are excluded. The seeds and the gelatinous material are then separated by one of the processes described below.

Seed extraction of large quantities of fruit

The extraction of seeds from large quantities of tomato fruit can be divided into two main types of operation. First, when the seed extraction is the only commercial operation involved, and second, when the seed is extracted as an additional product during the processing of tomatoes for purée or juice.

SEED EXTRACTION ON A COMMERCIAL SCALE   This system can be completely mechanized in that the fruit is harvested automatically before passing into a crusher. Alternatively, the fruit is hand-picked and either put directly into the crusher as it slowly passes through the field at a speed relative to the rate of hand-picking, or the picked fruit is transported to a stationary crusher in a seed-cleaning area.

The crusher squashes or crushes the fruit and the resulting mixture of gelatinous seed, juice and fruit residues is passed through a screen to separate off the gelatinous seed from the bulk of the remaining material. The crushed material is usually passed into a revolving cylindrical screen which allows the seeds and juice to pass through the mesh, while the fruit debris passes through the cylinder screen to drop in the field. Alternatively, the debris is collected separately for later disposal if the operation is stationary. The juice and seed mixture are each collected in a separate container. These machines are usually purpose-built by, or for, the seed-producing organization. The separated seed is then finally extracted from the gelatinous material and other materials by a separate process described later.

COMBINED JUICE AND SEED EXTRACTION   This system of seed extraction is done in collaboration with a tomato-processing factory. The factory line is generally organized to produce purée or juice processed for domestic use. During the operation, the purée or juice is separated from the relatively dry residual mixture of seeds, pulp and skins. Special lines or assemblies of apparatus are used in processing plants which intend to secure the seed in this way. One important feature is that the seeds are not subjected to the high temperatures used during tomato processing, and a purpose-designed modification of the normal processing plant is frequently referred to as a 'cold take-off'. The apparatus is normally made and sold by the specialist manufacturers of industrial tomato-processing equipment.

Close liaison must be maintained between the seed production unit and the tomato-processing organizations. One important feature is the need to have large batches of fruit of the same cultivar going through the factory and a thorough cleaning technique to avoid admixture of seeds from different cultivars. The system is used for the production of commercial quality seed for large-scale industrial tomato crops. It should not be used where seed of high genetic quality is required for further multiplication. Seed that has been extracted by this process has usually been separated from the mucilage during the industrial process and all that is required is washing to separate the seed from other debris.

When the tomato seed has been extracted from the fruit by one of the foregoing operations or systems, it is then usually necessary to separate the seed by a further wet upgrading method known as 'separation', before it can be finally washed and dried. The method of separation will depend on the quantity of seed to be processed, possible need to control a specific seed-borne pathogen or virus and the temperature of the local environment.

Seed separation

Separation by fermentation

The pulp containing the extracted tomato seed is left to ferment for up to 3 days at about 20–25 °C. But the rate of fermentation will depend on the ambient temperature and may even take up to 5 days. In warmer areas, such as California, the fermentation process is usually completed within 24 h. Frequent inspections will determine when the seeds' gelatinous coating has broken down. The mixture must be stirred several times a day to maintain a uniform rate of fermentation in the container and to avoid discoloration of the seed. It is usually necessary to cover the containers with muslin to reduce frit or fruit fly activity. The fermentation time must not be extended beyond that required for breaking down the mucilage or else the subsequent seed quality will be affected by premature germination.

There are claims that the fermentation process controls seed-borne bacterial canker of tomato (*Corynebacterium michiganense*), but the measure of success will depend on the duration of fermentation and the temperature reached.

Separation with sodium carbonate

This method is relatively safe and can be used for small quantities of seed in cooler temperate areas, where the fermentation method is not used. The pulp containing the extracted tomato seed is mixed with an equal volume of a 10% solution of sodium bicarbonate (washing soda). The mixture is left for up to 2 days at room temperature, after which time the seed is washed out in a sieve and subsequently dried. This method tends to darken the testa and is therefore not normally used for commercial seed lots, but is used by plant breeders and those who are involved in maintaining breeding material and parent lines.

Separation with hydrochloric acid

This method is favoured by the large commercial seed organizations as it produces a very bright clean seed sample. The actual dilution rate and duration of treatment used by the major commercial seed producers is usually confidential. This treatment is often combined with the later stages of fermentation. However, producers of relatively small quantities of tomato seed find that 567 ml of concentrated hydrochloric acid stirred into 10 l of seed and pulp mixture and left for half an hour is successful. It is very important that the acid is added to the water and pulp, and not the water and pulp to the acid, otherwise a dangerous effervescence will occur. All workers handling the concentrated and diluted acid solutions must wear appropriate face shields and protective clothing.

Control of seed-borne tobacco mosaic virus (TMV)

The first principle of minimizing the transmission of TMV via the tomato seed, in addition to the appropriate isolation distances from sources of infection, is to ensure that the fruits for seed extraction are only harvested from plants which do not show symptoms of infection. Where a relatively small number of plants is involved (which is especially possible where seeds of the determinate types are produced in greenhouses) fruit for seed extraction should be harvested only from the lower trusses of late-infected plants if healthy ones are not available (Broadbent, 1976).

There is evidence that both the sodium carbonate and the hydrochloric acid extraction methods will inactivate TMV transmitted on the testa of tomato seed (Nitzany, 1960). Another treatment, involving the use of trisodium orthophosphate, is used as an extra safeguard to inactivate high-value tomato seed. It is a separate treatment, carried out immediately following extraction, but before the seeds are dried. The seeds which have been extracted by one of the foregoing methods are soaked in 10% trisodium orthophosphate solution for 30 min and are then immediately rinsed in clear water and dried.

Seed washing

Tomato seeds which have been extracted by fermentation or acid treatment are immediately washed when the extraction time has been completed. This can be done on a small scale by washing in a series of sieves.

Large quantities of seed extracted from field crops are usually washed in long and narrow water troughs with a fall of 1 in 50 (see Fig. 12.3). The

Fig. 12.3. Washing extracted tomato seeds in a water trough.

trough has riffles at intervals and works on the principle that the seed is denser and sinks while the other debris floats off or goes through in suspension. The skill of doing this operation comes with experience and it is always a safeguard to have a suitably sized screen over the waste-water drain in case of mishaps.

Seed drying

This operation should commence immediately after the washing process has finished. Large-scale extraction units usually spin dry the seeds while they are in a stout cloth bag to remove excess water before the main drying.

In warm dry climates the seeds can be spread on hessian or other suitable mats and sun-dried, in which case the seeds are brushed to turn them occasionally. Large quantities of seeds are dried in rotary paddle hot air batch driers.

Seed yield and 1000 grain weight

Figures quoted for tomato seed yield vary significantly from one production area to another. The differences will vary according to several important features which include:

- Type of plant: determinate or indeterminate.
- Number of fruit trusses per plant: this can vary extensively depending on the training system or stage of 'stopping'.
- Plant density or population per unit area: this will be dictated, for example, by training system and irrigation system.

There are two alternative ways of estimating tomato seed yield:

1. According to weight of fruit.
2. According to unit area of plants.

In greenhouse production 1 kg of fruit will produce approximately 4 g of seed (approximately 1200 seeds). In field production one rule of thumb is that seed weight is 1% of the fruit weight, e.g. 1 t of fruit will produce 10 kg of seed. The expected tomato seed yield in the USA for field-grown determinate cultivars is between 250 and 400 kg of seed/ha. Workers in Africa report yields from 10 to 50 kg/ha.

The greenhouse tomato cultivars with a 1000 grain weight of 3.3 g tend to have larger seeds than the field or determinate types. The field or determinate types have a 1000 grain weight of 2.5 g.

Table 12.1. The main seed-borne pathogens of tomato species; these pathogens may also be transmitted to the crop by other vectors.

Pathogens	Common names
Alternaria alternata f. sp. *lycopersici*	
Alternaria solani	Early blight, Alternaria blight
Didymella lycopersici	Stem canker, stem rot
Fulvia fulva	Leaf mould
Fusarium oxysporum f. sp. *lycopersici*	Fusarium wilt
Glomerella cingulata	Anthracnose, ripe rot
Phoma destructiva	Fruit and stem rot
Phytophthora infestans	Late blight, fruit rot
Phytophthora nicotianae var. *parasitica*	Buck-eye rot
Rhizoctonia solani	Damping-off, collar rot
Verticillium dahliae	Verticillium wilt, sleepy wilt
Clavibacter michiganense	Bacterial canker, Grand Rapids disease
Pseudomonas syringae pv. *tomato*	Speck, bacterial leaf spot
Cucumber mosaic virus	
Potato spindle tuber viroid	
Tobacco mosaic virus	
Tobacco mosaic virus, streak strain	
Tobacco mosaic virus, tomato mosaic strain	
Tomato ringspot virus	

Pathogens

The main seed-borne pathogens of tomato are listed in Table 12.1.

Aubergine: *Solanum melongena* L.

Origins and types

This species has several widely used common names including aubergine, egg-plant and brinjal. The aubergine originates from India, where a wide range of wild types and landraces occur and is now generally grown as a vegetable throughout the tropical, subtropical and warm temperate areas of the world.

Aubergine is generally grown as an annual although it is a weak perennial. The plants usually develop some spines which are especially noticeable on older plants, but the degree of spininess can also depend on the cultivar.

The cultivars of *Solanum melongena* display a wide range of fruit shapes and colours, ranging from white, yellow, green, through degrees of purple pigmentation to almost black.

Cultivar description of aubergine

- There are open-pollinated and F1 hybrid cultivars.
- Plant attitude: erect, semi-erect or prostrate.
- Stem: anthocyanin coloration absent or present.

- Leaf: shape and margin.
- Fruit: curvature and shape, striped or single colour.
- Skin colour at commercial harvesting; white, yellow, green, mauve or purple.

(A detailed test guideline (117) is obtainable from UPOV; see Appendix 2.)

Flowering

The aubergine does not have a specific day-length requirement for flower initiation, but requires a higher minimum temperature than tomato. Optimum day temperatures for satisfactory growth and fruit production are around 25–35 °C. The aubergine is less tolerant to low temperatures than tomato, and cannot survive frost. If subjected to temperatures at the lower end of the range outlined above, the growth and development are very slow. However, aubergine tolerates humid tropical conditions better than tomato.

Pollination

The flowers of aubergine are normally self-pollinated, although some cross-pollination occurs when sufficient bees are attracted to the flowers.

There are F1 hybrid cultivars produced by specialist seed companies using male sterile lines as the female parent, or more commonly by emasculation followed by hand pollination. When male fertile female parents are used in the cross they are bud-pollinated.

The F1 hybrids of aubergine show some heterosis, but the greatest benefits claimed by commercial seed producers are earliness and tolerance to pathogens, e.g. TMV and fruit rot (*Phomopsis vexans*).

Plant raising and spacing

Aubergines are normally grown from a transplanted crop that has been raised in nursery beds. In some areas seedlings are raised in containers or seedbeds before pricking-off into nursery beds when the seedlings show their first true leaves. When seedlings are to be pricked out, the sowing rate is 60 g to 250 m² which provides sufficient plants to plant out 0.5 ha.

The optimum spacing of plants in their final positions is from 46 to 76 cm between plants in the rows and 60–120 m between the rows, depending on the vigour of the cultivar and any irrigation system used.

Rotation

Aubergines are susceptible to many of the soil-borne pests and diseases associated with other members of *Solanaceae* and crop rotations must take this into account. Generally a period of 4 years should elapse between successive aubergine crops or aubergines and other genera in *Solanaceae*.

Soil pH and nutrition

The soil pH should be approximately 6.5 and, if lower, sufficient lime should be added to bring it up to about 6.5.

Aubergines may be grown on raised beds in the humid tropics especially where occasional or regular heavy rainfall is expected, although on a large-scale seed production from plants grown in raised beds is unlikely to prove economic.

The crop responds to relatively high nutrient levels. Base dressings of 80–120 kg nitrogen, 80–100 kg P_2O_5, and 80–100 kg K_2O per hectare are given according to the soil nutrient status.

Nutritional studies made in India by Seth and Dhaudar (1970) have shown that the yield of aubergine seed increased at the relatively high levels of 70 kg/ha of nitrogen compared with lower nitrogen levels of 30 and 50 kg/ha. Although the N, P and K levels used by Seth and Dhaudar were lower than the general recommendations given earlier in this section, they demonstrate the value of nitrogen in increasing yields of aubergine seed. The nutrient levels used by Seth and Dhaudar are probably closer to the nutrient regimes used by small-scale farmers and seed producers in developing countries, whereas the higher recommendations are more in line with the large-scale farmers producing aubergine seed under contract by the more specialized producers.

The importance of nitrogen on accelerating flower formation in aubergine has been suggested by Eguchi *et al.* (1958) working in Japan.

Roguing stages and main characters to be considered

1. Before flowering: confirm desirable characters, including plant habit, pigmentation, vigour and foliage typical of the cultivar.
2. Early flowering and first fruit developing: confirm general plant habit, and characters as for stage 1, also the degree of spininess are according to the cultivar.
3. First fruit at market maturity: potential yield characters, fruit morphology, fruit colour, including stripes if according to cultivar description, spininess of calyx. If seed crop is for basic seed production, a sample of fruit should be examined for internal colour.

Production of F1 hybrid seed

F1 hybrid seed is produced by hand emasculation and pollen transfer is carried out as described for tomato. The flower buds and flowers are larger than those of tomato, and the work may therefore be regarded as less tedious. As with tomato, the male plants which will be used as a source of pollen are sown at a date to ensure that there is an adequate supply of pollen when the female line's anthesis commences. At the start of the pollinating programme, all new or old flowers on the female line which will not be used are removed.

It is also important to remove any developing fruit which have already set as a result of open pollination. The change of petal colour from white to a blue/violet colour is usually regarded as an indication that the anthers have already dehisced. Pollen collection from the male and transfer to the female parent follows the same procedure as for tomato. The cross-pollination of each flower is indicated with a secure but slightly loose tie around the individual flower's stalk (peduncle). The production of hybrid aubergine seed is described by Lal (1993a).

Harvesting

In order to ensure that seed development is complete, the fruits are usually hand-picked at a later or riper stage than for the market crop. Some seed producers leave the fruit on the plant until the abscission layer just behind the calyx is fully developed.

If hybrid seed is being produced, care must be taken to exclude fruit resulting from open-pollination, and in this case it is better to pick a fruit resulting from hybridization before it is likely to drop off the plant.

Seed extraction

There are two basic methods used for the extraction of aubergine seed, a wet and a dry extraction process. There is a general tendency to favour the wet extraction, especially in large-scale seed production. The dry process is still favoured where relatively small amounts of seeds are produced.

Wet extraction

For the primary extraction of aubergine seeds, the fruits are crushed and the seeds are separated from the remainder of the fruit pulp and debris as described for tomato seed extraction. But because aubergine fruit pulp is relatively dry, it is necessary to add extra water during and after fruit crushing in order to improve the separation of the seeds.

The seeds are extracted from the debris by the same process as tomato, but it is sometimes necessary to spray clear water into the separation screen cylinder to obtain the maximum seed yield. The extra water tends to wash the seeds free from the fruit pulp.

Dry extraction

In some countries the overripe aubergine fruits are dried in the sun until they shrivel. In the purple- and purple-black-fruited cultivars the drying out is accompanied by a fading of the skin colour to a coppery brown. The fruits are then hand-beaten, and the dried seeds hand-extracted.

This method is time consuming and laborious, but is used in some countries for the production of relatively small seed lots when ripe fruit are accumulated over several weeks and hand labour is available for the final extraction.

Table 12.2. The main seed-borne pathogens of aubergine; these pathogens may also be transmitted to the crop by other vectors.

Pathogens	Common names
Alternaria alternata	Leaf spot and fruit rot
Cochliobolus spicifer var. *melongenae*	
Colletotrichum melongena	Anthracnose
Fusarium oxysporum	Fusarium
Fusarium solani	
Phomopsis vexans	Fruit rot
Rhizoctonia solani	Damping-off
Sclerotinia sclerotiorum	
Verticillium albo-atrum	Verticillium wilt
Verticillium dahliae	
Eggplant mosaic	

Seed yield and 1000 grain weight

A satisfactory seed yield is approximately 150 kg/ha, but high-yielding cultivars grown in areas of good husbandry can yield up to 20 kg/ha.

The 1000 grain weight is approximately 5 g, but smaller seeded cultivars have a 1000 grain weight of approximately 4 g.

Pathogens

The main seed-borne pathogens of aubergine are listed in Table 12.2.

Peppers

The peppers that are grown as vegetables are *Capsicum* species, not to be confused with *Piper nigrum* L., which is in the botanical family *Piperaceae*, also commonly called 'pepper', the fruits of which are used as a spice or condiment, including the production of black pepper and white pepper.

The nomenclature of the genus *Capsicum* has been discussed by several workers, including Heiser and Smith (1953), Purseglove (1974), Andrews (1984) and Smith *et al.* (1987). The two *Capsicum* spp. which are generally considered important as vegetables are *Capsicum annuum* L. and *Capsicum frutescens* L.

C. annuum L. is usually an annual crop, with flowers and fruits borne singly. Cultivars of this species include the chillies, red, green, yellow and sweet peppers and paprika.

C. frutescens L. is a shrubby perennial with several flowers on each inflorescence; cultivated types include the bird chillies, cherry capsicum and the cluster pepper.

The *Capsicum* peppers originated in South America and spread into the new world tropics before subsequent introduction into Asia and Africa. They are now widely grown throughout the tropics, subtropics and warmer temperate regions of the world; they are not frost-tolerant. There has been an increase in the greenhouse and protected cropping production of the larger-fruited types in central and northern Europe for early- and out-of-season crops, and there has also been an increase in field, or partially protected crops the Mediterranean region for export. There is a wide range of fruit sizes and shapes in cultivation.

The physiology of the cultivated *Capsicum* peppers has been reviewed by Wien (1997a).

Cultivar description of sweet pepper

- There are open-pollinated and hybrid cultivars.
- Plant: general habit and branching.
- Leaf: length of blade and width.
- Flower: attitude, erect or non-erect.
- Fruit:
  - Colour before maturity: greenish-white, yellowish, green or purple.
  - Length.
  - Predominant shape.
  - Thickness of flesh: thin, medium or thick.
  - Capsaicin in placenta absent or present.
- Resistance to:
  - Tobamo-virus.
  - Potato Virus Y (PVY).
  - *Phytophthora capsici*.

(A detailed test guideline (012) for sweet pepper, hot pepper, paprika and chilli is obtainable from UPOV; see Appendix 2.)

Pollination

C. annuum and *C. frutescens* are generally self-pollinated, but some cross-pollination can occur between and within cultivars of the two species. Murthy and Murthy (1962) reported up to 68% cross-pollination in India. The high degree of cross-pollination experienced occasionally is due not only to large populations of pollinating insects, but also to dehiscence of anthers up to 2–3 days after individual flowers open. Prior to dehiscence, the stigmas are receptive to pollen transmitted from other plants.

For seed production and isolation purposes it should be assumed that all the types of cultivated peppers are cross-compatible rather than attempting to subdivide the genus into several species with different degrees of cross-compatibility.

Soil pH and nutrition

The pH of the soil should be between 6.0 and 6.5 and, if lower than this, a liming material should be added during site preparation.

Nutrient regimes with a relatively high level of nitrogen should not be used as they tend to delay fruit maturity. A general basic nutrient application during preparation is 115 kg/ha nitrogen, 200 kg/ha P_2O_5 and 200 kg/ha K_2O. However, the quantities of individual fertilizers applied should be modified according to the soil's nutrient status. Up to two-thirds of the nitrogen can be applied as top dressings once flowering and fruit set have commenced, especially in the humid tropics and areas where there is a significant leaching during crop establishment.

Irrigation

If the crop is to be grown in an area that is not rain-fed, an adequate supply of irrigation water must be available because the species are very prone to flower-drop; this and the dropping of undeveloped fruits occurs during times of water stress, more so with the larger-fruited types.

Field establishment

Peppers are either sown directly where they are to mature, or are raised in nursery beds and planted out as described for tomatoes. Raising the plants in nurseries ensures that the seedlings can be protected from late frosts, heavy rain or wind.

Final plant density depends on both the vigour of the cultivar and the irrigation system to be used. Generally the final plant stand is 30–60 cm within the rows and 45–90 cm between the rows.

Roguing stages and main characters to be observed

1. Before flowering: check the desirable characters typical of the cultivar, including growth habit, vigour, foliage, pigmentation and leaf morphology.
2. Early flowering and first mature fruit: check characters as for stage 1, in addition pay particular attention to presence of any seed-borne diseases.
3. First fruit at market maturity: check characters as for stages 1 and 2, also check fruit colour and morphology.

Basic seed production

Crops for basic seed production are rogued, or plants selected, according to the criteria listed under cultivar description and roguing stages. Increased pungency can be introduced into sweet or mildly pungent types by pollen contamination from types with a higher pungency level, and it is therefore important to check that the absence of or degree of pungency is according to the cultivar type. This can be achieved by examining and tasting a small piece of the fruit's placenta-wall tissue from each selected plant. This process is not necessary for the hot chilli types, but it is important that the stock of a mild type does not increase in pungency. The internal fruit characters to be examined for basic seed also include thickness of fruit wall and internal fruit flesh.

Production of F1 hybrid seed

The technique for hybrid seed production is similar to that which has been described for tomato and aubergine. It is important to ensure that flower buds in which the anthers have already dehisced are not used. Poulos (1993) has described the maintenance breeding and hybrid seed production of peppers.

Seed extraction

There are two methods of pepper seed extraction, a wet and a dry. It is not possible to completely dry the large fleshy fruited types, and they are therefore usually wet-extracted without fermentation. The crushing and wet extraction process is similar to that described for aubergine.

The small-fruited pungent types can be successfully dried in the sun before seed extraction, or put into batch driers until shrivelled and dry. In areas where there is sufficient hand labour, the dried fruits are hand-flailed to extract the seeds but elsewhere they are put through a thresher. Further separation is done by winnowing or in an aspirated screen cleaner. Hand-extraction of seed of pungent types is not pleasant for workers, as they are very likely to experience irritation to eyes and mucous membranes.

Where there is large-scale production of peppers for a dehydration plant, good-quality seeds can be recovered by washing the separated seed from associated materials. Figure 12.4 illustrates the harvesting of pepper fruits in California, USA, where the seed was to be extracted in collaboration with a dehydration factory. The seed can be separated as described for tomatoes by a water trough (flume) process and then dried. Further upgrading of the clean sample can be achieved by passing it over an aspirated screen cleaner after drying. Close liaison should be maintained with the processing plant to ensure a high degree of trueness-to-type and no adverse high temperature effect on the seed.

Seed yield and 1000 grain weight

The seed yield in any one area will usually depend on whether the cultivar is a pungent or sweet type. The pungent types are generally the higher yielding of the two. A satisfactory seed yield is from 100 to 200 kg/ha.

Fig. 12.4. Harvesting pepper fruits in California, USA, for seed extraction in conjunction with a dehydration factory.

A useful rule of thumb for small quantities of fruit is that 1 kg of small-fruited pungent types will yield 25–100 g of seed, while 1 kg of sweet or large-fruited types will yield 5–50 g of seed.

The 1000 grain weight is *c.*3.5 g for the pungent types and 5 g in the sweet types.

Pathogens

The main seed-borne pathogens of *Capsicum* species with common names are given in Table 12.3.

Potato, European Potato, Irish Potato, White Potato: *Solanum tuberosum* L.

The alternative common names for this species distinguish it from the sweet potato (*Ipomoea batatus* (L.) Lam.), which is in the botanical family *Convolvulaceae*. *Solanum tuberosum*, which originates from South America, is widely grown in the temperate regions of the world and is increasing in popularity in the subtropics and tropics.

The commercial and home garden crops have traditionally been produced vegetatively from tubers, generally referred to as 'seed potatoes' with plant breeders the main workers producing plants from true botanical seed. Reports by the International Potato Center (CIP, 1979) have indicated an interest in the

Table 12.3. The main seed-borne pathogens of *Capsicum* species; these pathogens may also be transmitted to the crop by other vectors.

Pathogens	Common names
Alternaria spp.	Fruit rot
Cercospora capsici	Frog-eye leaf spot, fruit stem-end rot
Colletotrichum capsici and	
Colletotrichum piperatum	Ripe rot, anthracnose
Diaporthe phaseolorum	Fruit rot
Fusarium solani	Fusarium wilt
Gibberella fujikuroi	
Phaeoramularia capsicicola	Leaf mould, leaf spot
Phoma destructiva	Phoma rot
Phytophthora capsici	Phytophthora blight, fruit rot
Rhizoctonia solani	Rhizoctonia
Sclerotinia sclerotiorum	Sclerotium rot, pink joint, stem canker
Pseudomonas solanacearum	Brown rot bacterial wilt
Xanthomonas campestris	Bacterial spot of fruit, stem and
pv. *vesicatoria*	seedling blight
Alfalfa mosaic virus	
Cucumber mosaic virus	
Tobacco mosaic virus	

production of commercial potato crops from seed, especially for developing countries. The term now widely accepted for seed of this crop is 'true seed', often referred to as 'TPS', which distinguishes it from tubers.

Several advantages are cited for the promotion of the use of TPS, which can be of significance, and with important application in developing countries. These include the main fact that if using TPS for the production of the next potato crop, all tubers can be used for food supply, thus retaining a valuable food source. In addition, TPS saves on importation and storage of bulky planting material, and avoids seed tuber losses resulting from storage diseases and other storage hazards. Jones (1982) reported that some of the economically important potato viruses are not seed-transmitted.

Cultivar description of potato

- Season of use: suitability for specific climatic regions, day-length and market outlets.
- Tuber characteristics: shape, external skin colour, internal colour, characters of 'eye', including pigmentation.
- Foliage ('haulm'): relative height.
- Flower: colour.
- Resistance to specific pathogens.

(A detailed test guideline (023) is available from UPOV; see Appendix 2.)

Agronomy

True potato seed (TPS) systems can be used for the production of ware potatoes or for the production of seedling tubers for a following crop or crops.
Umaerus (1989) has reviewed the potential of TPS and its applications.

Seed extraction

The method of seed extraction widely adopted was described by Sadik (1982).
The macerated berries are placed in a modified funnel similar to the apparatus
used by nematologists to extract nematode cysts from soil suspensions. The
apparatus is basically a funnel with a cylinder attached and a rubber bung, with
a coiled copper tubing fitted into the neck of the funnel. The macerated berries are placed in the top and as water passes through the coiled tubing, it
creates a cyclonic agitation. The scum and fruit debris overflow at the top
while the seeds collect at the bottom of the apparatus and are retrieved by
removing the bung.

Following this wet extraction process, the seeds are dried at room
temperature.

Seed yield and 1000 grain weight

Accatino and Malagamba (1982) reported that an average flowering plant produced approximately 20 berries, each containing *c*.200 seeds.

The 1000 grain weight is 0.6 g.

Pathogens

The main seed-borne pathogens of potato are listed in Table 12.4.

Table 12.4. The main seed-borne pathogens of potato; these
pathogens may also be transmitted to the crop by other vectors.

Pathogens	Common names
Erwinia spp.	Blackleg, soft rots
Potato (Andean) latent virus	
Potato black ring virus	
Potato black ringspot virus	
Potato spindle tuber viroid	
Potato virus T	
Potato virus X	
Potato virus Y (Potato virus X and Potato virus Y together cause 'rugose mosaic')	
Tobacco ringspot virus	

Further Reading

Bosland, P. and Votava, E. (1999) *Peppers: Vegetable and Spice Capsicums*. CAB International, Wallingford, UK.

Heuvelink, E. (2005) *Tomatoes*. CAB International, Wallingford, UK.

Lal, G. (1993) Seed production techniques of tomato and brinjal in the tropics and subtropics. In: *Breeding of Solanaceous and Cole Crops*. AVRDC, Taiwan, pp. 61–81.

13 *Apiaceae* (formerly *Umbelliferae*)

There are many genera in this botanical family which are cultivated as vegetables for cooking, salads, condiments or flavouring. The main vegetables whose seed production is discussed below are:

Daucus carota L. sub sp. *sativus* (Hoffm.) Thell.	Carrot
Pastinaca sativa L.	Parsnip
Petroselinum crispum (Mill.) Nyman ex A.W. Hill	Parsley
Apium graveolens L. var. *dulce* Mill.) DC. syn. *Apium dulce* Mill.	Celery
Apium graveolens L. var. *rapaceum* Mill.) DC.	Celeriac, root celery, turnip rooted celery

These five vegetable crops are extremely important in temperate regions of the world. Celery and carrot are especially popular in North America and Europe but they are also important locally in other areas of the world including some subtropical regions.

Other crops in *Apiaceae* are either of local importance as vegetables or salads, or are used as flavourings, garnishes or condiments. The following crops are usually grown from seed:

Anethum graveolens L.	Dill
Anethum sowa Kurz.	Indian dill
Anthriscus sylvestris Hoffm.	Chervil
Carum carvi L.	Caraway
Coriandrum sativum L.	Coriander
Cuminum cyminum L.	Cumin
Foeniculum vulgare Mill.	Fennel
Foeniculum vulgare Mill. var. *dulce* (Mill.) Thell.	Florence fennel
Pimpinella anisum L.	Anise

Some species in this family are important sources of essential oils, especially the dills, celery, caraway and cumin.

Carrot: *Daucus carota* L. ssp. *sativus* (Hoffm.) Thell.

Origins and types

The modern cultivated carrot has been derived from the wild carrot *Daucus carota* L. found in Europe, Asia and Africa.

There appear to be two main sources of material from which early selections were made; these are the anthocyanin carrots from Asia (especially centred in Afghanistan where purple-rooted carrots may still be found in cultivation) and the carotene carrots developed in Europe. Banga (1963) described the descent and development of the main types of the 'Western carotene carrot'.

Carrot cultivars are usually grouped according to maturity period, root shape (length and width), root-tip shape and foliage. Further divisions can be made according to their season of use and root colour. Banga (1964) has described the types of orange-rooted carrots and their specific characters. The six recognized root types of carrot are shown in Fig. 13.1.

Research and current trends in carrot seed production

The individual carrot flowers, in common with most other species in *Apiaceae*, are born on terminal branches in compound umbels. There is a distinct order

Fig. 13.1. Types of carrot roots, left to right: short horn, Nantes, Amsterdam forcing, Chantenay, Autumn king and St Valery.

of flowering which relates to umbel position. The first umbel to flower is the primary (sometimes referred to as the 'king') umbel which is terminal to the main stalk. Branches from the main stalk form secondary umbels, and subsequent branches from these form tertiary umbels. Quaternary branches and umbels may also be formed.

The modern methods of carrot-root crop production for the fresh market and for processing require high-quality carrots with the minimum of variation between individual roots derived from the same seed lot. The main sources of variation in carrots and the implications for changes in seed production techniques have been reviewed by Gray and Steckel (1980), who later reported on the influence of flowering time in the seed crop on seedling size in the subsequent carrot crop (Gray and Steckel, 1985b). The relationships between seed weight and endosperm characteristics were discussed by Gray and Ward (1985). The implications of variation in seed weight and embryonic weight in the root vegetables, including carrot, have been more recently reviewed by Benjamin *et al.* (1997). Variation in weight of individual seeds and the weight of their embryo can contribute to the subsequent variability of seedling size. Because the volume of embryo in the carrot seed contributes to less than 5% of the seed's total volume, the variation within a given seed lot's grading alone does not produce a useful improvement of a seed lot. Earlier work by Gray (1979) established that there is a clear relationship between umbel position and seed quality.

Plant population and seed yield

Carrot seed yield increases with plant population. In the 1950s, Hawthorn (1951) and Hawthorn and Pollard (1954) reported that average seed yields per annum increased from 580 kg/ha to 1039 kg/ha, as the number of stecklings planted increased from 12,000 to 140,000/ha. More recent work by Gray (1981) extended the concept as suggested by earlier workers that an even greater seed yield could be obtained by further increasing the plant population per unit area. Gray investigated the possibility of increased plant densities in root to seed method from 100,000 to 800,000 stecklings/ha, and from 110,000 to 2,560,000 plants/ha by the seed to seed. Seed yields increased from 1250 to 2000 kg/ha for the root-to-seed method and from 700 to 2400 kg/ha for the seed-to-seed method (see Fig. 13.2).

The high plant densities also improve seed quality because there is less branching and therefore a higher proportion of primary to secondary umbels than at the lower plant densities. This is illustrated in Fig. 13.3. A further advantage is that overall seed quality is improved by a shorter timespan of seed maturity between the first and last umbels of each plant.

In practice, increasing plant densities in mechanized systems depends on the ability of transplanting machines to plant stecklings at the higher densities. The main advantage of the higher plant densities is the shortening of the overall flowering period, increased evenness of umbel ripening and a higher proportion of seed in the final seed lot derived from primary umbels, thus contributing to improvement of seed quality. A uniform carrot seed crop

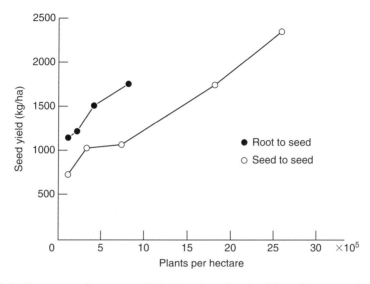

Fig. 13.2. Response of carrot seed yield to plant density. (From Gray, 1981.)

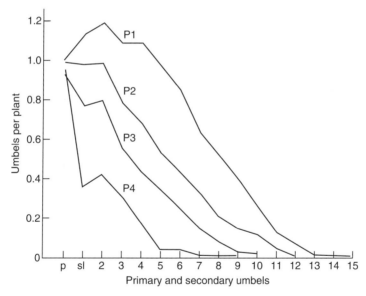

Fig. 13.3. Effect of carrot plant density on branching in the seed crop. P1, P2, P3, P4 = 100,000, 200,000, 400,000 and 800,000 plants per hectare, respectively. (From Gray, 1981.)

facilitates the application and timing of pre-harvest desiccant sprays or polyvinyl acetate (PVA) adhesives where the weather conditions dictate their use. Other advantages include easier management of the timing of harvests to secure the maximum seed yield.

Temperature effects on seed quality

Investigations into the effects of day and night temperature fluctuations on the mother plant of subsequent seed quality were made by Gray *et al.* (1988a) over a 2-year period. They found that in the cultivar Red-cored Royal Chantennay an increase of day/night temperature from 20/10 °C to 30/20 °C reduced the mean weight per seed by 20% in 1984 and 13% in 1985. They did not find any temperature effects on endosperm, endosperm cell number or embryo weight. However, they found that the pericarp weight decreased with the increase of temperature. The largest embryos, highest nitrogen, DNA and RNA contents were in the seeds produced at the highest temperatures; these seeds germinated and emerged earlier and had a higher percentage of seedling emergence than the seeds produced at the lowest temperatures.

Influence of seed position on mother plant on seed quality
and seedling performance

Germination and seedling emergence evaluations were made by Thomas *et al.* (1978) with carrot seed harvested separately from primary and secondary umbels when either immature (i.e. 47 days after anthesis) or mature (i.e. 68 days after anthesis). Germination was approximately 20% lower in the immature seeds at temperatures below 12 °C and greater than 25 °C. Immature seeds from secondary umbels took approximately 6 days longer to germinate at 5 °C than those from primary umbels. In addition, the weights of seedlings from primary umbel seeds were greater than those from secondary umbels when weighed at about 20 days after 50% emergence.

This interesting result was substantiated by Gray (1979) who showed clearly the relationship between umbel position and seed quality (Fig. 13.4). This work has resulted in the adoption of higher plant densities in commercial carrot seed production. Because primary umbel seed is ready for harvest some time in advance of umbels of a lower order, especially at low plant densities, hand-harvesting operations have been used in some areas of the world to secure the seed from the primary umbels. In mechanized production systems, changes have taken place in planting systems to increase plant density.

Cultivar description of carrot

- There are both open-pollinated and hybrid cultivars.
- Season: early (suitable for protected cropping) or main crop.
- Root type: based on shape, relative size, length and root tip (see Fig. 13.1).
- Market use: bunching, pre-packing, processing, canning, freezing or dehydration (high dry-matter content for dehydration), storage or remain in ground.
- Foliage: amount of foliage, height, petiole base colour.
- Root: length, type, shape of shoulder and root tip. Root surface texture, colour, core colour for sample of crop.

(A detailed test guideline (049) is obtainable from UPOV; see Appendix 2.)

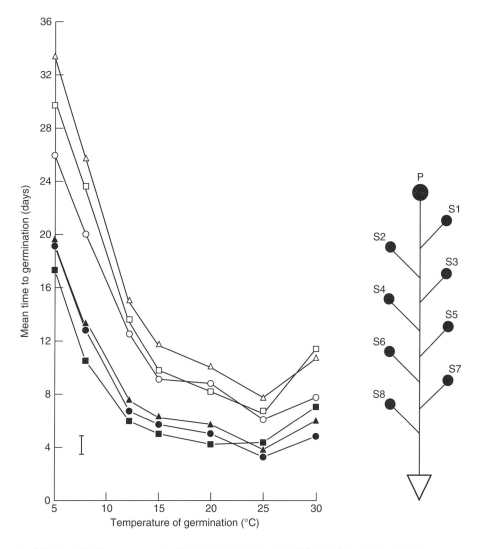

○ Primary umbels   ● From crops harvested on 10/7 and for primary umbels
□ Secondary 1 umbels   ■ Secondary 1 umbels
△ Secondary 8 umbels   ▲ Secondary 8 umbels from crops harvested on 7/8;
                         I = L.S.D. ($p = 0.05$).

Fig. 13.4. Relationship between mean time to germination and temperature for different umbel positions. (From Gray, 1979.)

Flowering

The carrot cultivars that originated in Asia tend to be annuals when grown in long days, and do not have a vernalization requirement.

The cultivars that have been developed in the temperate regions of Europe and North America are biennials which normally produce a swollen taproot at the end of the first season, with a rosette of leaves. The plant generally requires a cold period before it will bolt. The response to cold is not necessarily related to root size. Even in the types that require a cold period for flower initiation, there are likely to be some plants that have the annual habit; these plants which bolt during their first year also have poor root quality and are identified and rejected during the course of routine roguing.

Atherton *et al.* (1984, 1990) and Craigon *et al.* (1990) have made detailed studies of the prediction, temperature and effects of photoperiod on flowering in carrot.

Pollination and pollinating insects

Individual carrot flowers are normally protandrous and much cross-pollination occurs between plants in a seed crop. However, because of the extended flowering period, resulting from several successive umbels per plant and the succession of flowers on individual umbels, the possibility of self-pollination always remains.

Bohart and Nye (1960) observed the occurrence of pollinating insects on flowering carrots in Utah, and noted that while honeybees were efficient polli-nators, they were frequently scarce on carrot crops because other crop species were flowering in the vicinity at the same time. They also observed that several insect genera in Hymenoptera, Diptera and Coleoptera were extremely import-ant pollinators of carrot seed crops, in the absence of bees.

Hawthorn *et al.* (1960) found that when there was an adequate supply of pollinating insects, both seed yield and seed quality were high; where natural insect pollinator populations were low, there was an advantage in increasing the number of honeybee colonies by placing groups of hives adjacent to the carrot fields.

Isolation

Because of the high possibility of cross-pollination, isolation distances for com-mercial seed crops should be 1000 m. For basic seed the distance should be at least 1600 m.

In areas which specialize in carrot seed production, the different cultivars within the same type can be zoned; this minimizes cross-pollination between the different root types.

Cultivated carrots cross-pollinate very readily with the wild carrot, and this must be taken into account when choosing sites for seed production. Contamination of seed crops by wild carrot pollen is a major reason for genetic deterioration of seed stocks in some areas of the world.

Soil pH and nutrition

The optimum soil pH is 6.5; soils with a pH lower than this should either be avoided or adequate liming materials applied during site preparation as the car-rot crop will not tolerate acid conditions.

Little experimental work has been done on the nutrition of carrots for seed production. Most of the commercial seed producers rely on either personal observations or recommendations based on the production of marketable root crops when preparing soils for the first season of steckling production. These are usually N:P:K ratios of approximately 1:2:2, with the phosphorus application being lowered for soils with a high phosphorus status.

Experimental work by Austin and Longden (1966) indicated that improved seed quality resulted from increased soil fertility achieved by inorganic fertilizers and organic manures applied prior to the year of flowering.

Cooke (1982) suggested that there is a clear case for supplying sodium to carrots for root production, and reported that in the UK 380 kg/ha of sodium chloride increased yields to the same extent as 190 kg/ha of potassium chloride.

Top dressings of up to 200 kg/ha of nitrogen are applied in the spring of the second year by some carrot seed producers. Nitrogen applied in this way will counteract losses from leaching especially in areas of high winter rainfall.

Agronomy

There are two systems of carrot seed production, 'seed to seed' and 'root to seed'.

Seed-to-seed production
In this method, the stock seed is sown in the late summer (July or August in the northern hemisphere) and grown *in situ* through the following winter. The plants flower the following spring and seed is harvested in the late summer of the year following sowing.

Row spacings of 50–90 cm are used, with a sowing rate of 2–3 kg/ha. The roots cannot be inspected or rogued, and in practice virtually no roguing is done for foliage characters either; the system relies therefore on very good quality stock seed and satisfactory isolation.

Root-to-seed production
This system is similar to the seed-to-seed system as far as timings are concerned, but the plants ('stecklings') are raised in beds and transplanted in the spring. Depending on local customs and winter conditions, the stecklings are either left *in situ* during the winter or lifted in the late autumn and stored until replanting in the spring.

The ratio of steckling bed to final area transplanted is from 1:5 to 1:10, depending on the degree of roguing, if any, prior to storage or transplanting into final quarters and also the degree of uniformity of transplant used.

The transplanted steckling rows are 75–90 cm apart with 20–30 cm between plants. The sowing rate is 6–8 kg/ha for the root bed. Raising stecklings that are later transplanted from their beds offers the opportunity of roguing plants with undesirable root or foliage characters when lifting and planting, but the system is very labour-intensive and is declining. One advantage is that the area of land used in the first year is significantly less than the area subsequently planted for seed production.

A variation of this method involves the production of mature roots as for fresh markets. The roots are lifted in late winter. Some growers mow off the carrot tops before lifting, but care must be taken not to damage the root crowns. About 5–8 cm of the leaf tissue is left on the roots. In some areas the roots are replanted without storage. In this case, it may be necessary to protect the transplanted roots from frost by applying a layer of litter or straw to the surface of the beds.

It is important that any diseased, misshapen or mechanically damaged roots are discarded during the roguing of the lifted roots. Roots should be relatively free from soil, but not washed, before storage. The method of storage depends to some extent on local traditions. In Scandinavia, straw- and soil-covered pits or 'clamps' are used but in other areas barns or purpose-built stores are used. Fine soil, moist peat or sand is placed between layers of roots.

The ideal storage temperature is 1 °C with humidity at about 90–95%. Controlled ventilation stores use outside air for cooling when the ambient temperature is above freezing.

Planting out

Roots (or stecklings) are planted out from the store in the early spring when soil conditions are satisfactory. The roots must not be allowed to dry out at this stage as this adversely affects establishment and subsequent seed yield.

Large-scale seed producers use planting machines, whereas dibbers, trowels or spades are used on a small scale. Whatever planting system is adopted, the crown of the root must be at, or just below, the final surface of the firmed soil. The post-planting time is critical for quick plant establishment, and if conditions are dry, irrigation should be applied. It is useful in some areas to earth up with about 10 cm of soil when the plants are re-established.

Roguing stages and main characters to observe

Seed-to-seed production

Very little, if any, roguing can be done when the crop is grown without lifting. However, plants bolting early and those with atypical foliage characters should be removed. If the crop is lifted and replanted, it is rogued as described below for root-to-seed system but very little confirmation of root type can be done.

Root-to-seed production

During the first year's growing season:

1. Remove plants displaying atypical foliage. Remove plants bolting in the first year.

2. After the roots have been lifted inspect for trueness to type, according to root shape, colour and size. Discard roots showing poor colour, incorrect colour, off-coloured shoulders (purple, green), split or fanged roots or those with rough surfaces.

Harvesting seed

In small-scale commercial production and where hand labour is plentiful, the umbels are cut by hand as they ripen. This system is also used for small plots of high-value seed.

On a larger scale, where mechanized systems are used, it is necessary to decide the best cutting time for maximum seed yield. There is a tendency for some carrot seed to be lost by shattering or dropping from the primary umbels if cutting is delayed to allow maturity of seed on secondary and tertiary umbels. In some areas of the world the plants are sprayed with PVA to prevent loss from the primary umbels. Harvesting early in the day when the dew is still on the seed heads will also reduce the amount of seed which is lost from the umbels.

Generally, the crop is cut when the earliest maturing seed on the primary umbel is starting to drop. Carrot seed is brown when ripe, and at this stage the umbel is brittle. The cut crop continues to dry in windrows. High plant densities tend to produce a concentration of umbels in the upper part of the individual plant stalks; thus, if the cutter bar can be raised during cutting, the subsequent windrows rest on longer stubble, allowing increased air circulation under them prior to threshing.

When the seed crop has ripened, it can be separated from the remainder of the plant debris by a combine harvester. As with the initial cutting, the combining can be done early in the day to take advantage of dew reducing the loss from dropping. The plant material must not be too dry or, in addition to losing seed, the plant debris will shatter and increase the subsequent seed cleaning operation. Where the scale of operation is inappropriate for combine harvesters, the dry material is put through a thresher.

The effects of threshing and conditioning (drying) carrot seed harvested at different stages of development and dried at a range of temperatures was investigated by Tucker and Gray (1986). They found that the seed which gave the highest percentage germination, most rapid germination and seedling emergence was derived from seed crops harvested from 50 days after anthesis and conditioned within the temperature range of 15–35 °C. They also found that removing the seed from the umbel head at harvest but before drying had no effect on subsequent seed and seedling performance compared with seed conditioned in the umbel head.

Carrot (similar to dill and caraway) has spines or 'beards' on the seed (see Fig. 13.5). The seeds' spines must be removed by a debearder before further cleaning operations. Debearding improves the seed flow and reduces the volume of the seed lot. Further cleaning is achieved by aspirated screens and indent cylinders.

Seed yield and 1000 grain weight

The yield of carrot seed at different plant densities is the subject of current investigations in several research centres of the world. At present expected

Fig. 13.5. Carrot seed sample with spines ('beards') (left) and after processing in a debearder.

yield of open-pollinated cultivars in the temperate region is about 600 kg/ha with highest yields achieved reaching 1000 kg/ha.

Yields in the tropical regions are usually considerably lower despite using higher altitudes to achieve satisfactory vernalization, and figures of about 300 kg/ha should be expected for the European types. The Asiatic types produce only about 250 kg/ha when seeded in the tropics.

The 1000 grain weight is about 0.8 g.

Production of F1 hybrid seed

Hybrid cultivars of carrot have the advantage of relatively uniform roots and have been produced by the use of two methods, both are described by Riggs (1987). The first system is with cytoplasmic male sterility (CMS) in which the pollen does not develop beyond the microspore stage, sometimes referred to as the 'brown anther form'. The other type is the petaloid form in which the five anthers are transformed into petaloid structures during their early development and do not produce any pollen. The final morphology of the petaloid anthers varies from petal-like to filamentous (Eisa and Wallace, 1969). According to Riggs, most commercial F1 carrot hybrids are produced from the petaloid CMS. Generally the ratio of female-to-pollinator rows is from 2:1 to 4:1 (Takahashi, 1987). Production of parental lines and the hybrid seed by either system should be carried out in accordance with the instructions of the maintenance breeder.

Hybrid seed yields can be relatively low, about 500 kg/ha, after taking into account the exclusion of the seed from pollinator rows. The lower yield is frequently attributed to less insect activity because of smaller petals on the male sterile flowers of the seed-producing lines. Seed producers frequently use colonies of bees to supplement the natural level of pollinating insect activity when producing hybrid carrot seed.

Improvement of basic stock of open-pollinated cultivars

The following points are made to assist the improvement of basic seed stocks of open-pollinated cultivars:

1. Grow the crop on a root-to-seed system.
2. Select roots for further seed production to a very high standard of trueness to type and freedom from defects before storage or replanting.
3. Inspect selected roots again at the end of storage, especially to reject roots showing storage diseases.
4. When selecting to improve colour of root core, check selected roots immediately before replanting by either cutting a small portion off the root tip, or take a small transverse core with a cork borer. Treat the cut surface with a suitable fungicide powder and plant immediately.
5. For stock seed all plants can be harvested; progeny testing can be done before bulking.

Pathogens

The main seed-borne pathogens of carrot are listed in Table 13.1.

Table 13.1. The main seed-borne pest and pathogens of carrot species; these pathogens may also be transmitted to the crop by other vectors.

Pathogens	Common names
Alternaria dauci	Carrot leaf blight
Alternaria radicina	Black root rot, seedling blight
Cercospora carotae	Cercospora blight of carrot, leaf spot
Erysiphe heraclei	Powdery mildew
Phoma rostrupii	Phoma root rot
Xanthomonas campestris pv. *carotae*	Bacterial blight, root scab
Carrot motley dwarf caused by a complex of three viruses, including carrot red leaf virus	
Ditylenchus dipsaci	Eelworm

Parsnip: *Pastinaca sativa*

Origins and types

Parsnips are essentially a European vegetable but are grown as a relatively minor crop in North America and have also been introduced to other temperate parts of the world.

Traditionally, the root is used for culinary purposes in the autumn and winter at the end of the first season. However, with modern pre-packaged marketing, harvesting of smaller roots takes place over a longer period, commencing in the summer with roots produced at close spacings.

There is a relatively small range of root types represented in the available parsnip cultivars. The basic shapes are wedge, bayonet and bulbous. In recent years, plant breeders have put emphasis on the production of cultivars which are resistant to canker.

The parsnip plant is a biennial which produces a swollen taproot with a rosette of leaves. The root is fully developed by the end of the first season when the leaves die down. It flowers in its second year following natural vernalization of the roots during winter. No research work has been reported on the low temperature requirement; in northern Europe, the dormant roots of parsnip have always vernalized by the end of winter and new leaves are produced from the crown of the root by early spring.

The physiology of parsnip and some other root vegetables have been reviewed by Benjamin *et al.* (1997).

Cultivar description of parsnip

There are open-pollinated and F1 hybrid cultivars. The cultivar descriptions of parsnip are normally based on root characters.

*   Use: season, suitability for processing (canning), fresh market, pre-packaging.
*   Root:
    ○   Shape bulbous, wedge, bayonet or intermediate.
    ○   Relative length.
    ○   Crown shallow or hollow (deep).
    ○   Shoulder square or round.
    ○   Surface texture and colour; internal flesh colour, white or cream.
*   Resistance to canker, bruising during market transit and incidence of fanged roots are usually included in varietal characters.

(A detailed test guideline (218) is obtainable from UPOV; see Appendix 2.)

Agronomy

The roots of parsnip are not sensitive to frost, and it is therefore possible to produce a seed crop by the seed-to-seed method without the need to lift, store and replant the roots. However, for basic seed and other high genetic quality stock seed production, it is vital that the root-to-seed method is used so that the choice of plants for seeding is based on root characters.

The parsnip root weight and size are usually directly related to the length of time it is growing in its first year. Therefore, in order to make selections of roots in the autumn which are typical of the cultivar, it is necessary to sow early in the spring.

Germination and seedling emergence can take up to 3 weeks and this is another factor contributing to the need for relatively early sowing as soon as soil conditions are suitable for the root-to-seed method.

Soil pH and nutrition

The best quality roots are produced on deep peat, silt or sandy loam soils. The presence of stones will cause forking or fanged roots; shallow or stony soils should be avoided if roots are to be lifted for selection. This crop requires a soil with a pH around 6.5; it will not succeed when the pH is less than 6.0.

A fertilizer with a ratio of 1:2:2 of N:P:K is applied during the final preparation of the seedbed, according to the nutrient status of the soil.

Many parsnip seed producers apply a top dressing of a nitrogenous fertilizer in the spring following vernalization once the plants are seen to be producing leaves. This can be applied as ammonium sulfate at a rate of 200 kg/ha. The top dressing is especially useful in areas where there is a high leaching rate during the winter or early spring.

Plant production

Seed-to-seed method
Seed is sown in the Northern Hemisphere in early summer, usually between mid-May and mid-June in rows 1 m apart. At this row spacing the sowing rate is 4 kg/ha. The later sowings may run into moisture-deficit problems for satisfactory germination. In North America, irrigation is usually necessary to ensure satisfactory seedling emergence. Because of the relatively slow germination the seed should either sown in a 'stale seedbed' or a suitable pre-emergence residual herbicide applied.

The seedlings are thinned to a distance of 60 cm if the plants are to flower in situ; if they are to be lifted for root inspection, then half this distance in the row is satisfactory, but they are replanted at the greater distance after selection.

Root-to-seed method
The same system is followed as for the seed-to-seed method, except that if stock seed is being produced, the sowing date should be in accord with the sowing date for the fresh market crop. This will ensure that the roots have

reached their typical size and shape as well as showing other characters when they are lifted for selection in the autumn.

The sowing rate is similar to the seed-to-seed rate, but growers who adopt a closer spacing use an increased sowing rate. This will depend on the machinery used for lifting the roots.

The roots are lifted in the late autumn or around mid-winter according to local custom based on whether or not the ground is regularly frozen in the middle of winter.

The method used for lifting roots depends on the scale of operation. In large-scale production they are ploughed out, but on a smaller scale, especially for production of basic seed, the roots are forked out. The roots are examined for trueness-to-type as soon after lifting as possible and must not be allowed to dry out.

After selection, the roots to be retained for seed production the following season are replanted, usually in rows 1 m apart with 30–60 cm between plants within the rows. If the ground is frozen, small quantities of roots selected for basic or breeders' seed can be stored in moist soil, sand or peat until planted.

Pollination

Parsnip is predominantly cross pollinated. The flowers are protandrous, which reduces the chance of selfing of individual plants, but some selfing occurs. There is a tendency for the last flowers on the ultimate umbels not to be fertilized. (This must be taken into account when choosing the cutting time.)

Pollination is by a wide range of insect species including *Diptera* and *Lepidoptera*, but bees are the main pollinators. Seed producers frequently supplement natural insect populations with colonies of hive bees. The honey produced by bees working on parsnip flowers is reputed to be dark in colour and of very good quality.

Small enclosed lots of flowering parsnips in cages or structures are generally pollinated by blowflies put into the enclosure at regular intervals according to their lifespan.

Isolation distances should be a minimum of 500 m between parsnip seed crops, including commercial root crops which may contain a small percentage of bolters in their first year.

Roguing and selection of parsnip

The roguing or selection of parsnip is normally only done when the roots have been lifted. Thus, the seed-to-seed crops are not normally inspected. Plants showing atypical characters should be discarded. Plants which bolt in their first year and weeds in *Apiaceae* are removed during roguing.

Parsnip root characters to be observed when selecting are:

- Shape: relative length/width, shape of shoulders.
- Crown: degree of depression.
- Quality incidence of fanged or forked roots, although this may have been environmentally induced, forked and split roots should be rejected.
- Colour: white or cream, according to cultivar.
- Surface texture: smooth.
- Resistance to pathogens responsible for canker, i.e. *Itersonilia pastinacea, Phoma* spp. and *Centrospora acerina*.
- Bolting plants: uniformity of height and colour of individual plants.

Harvesting seed

Parsnip plants grown for seed production are very tall, and the mature plant with umbels can reach more than 2 m in height. Crops produced by the seed-to-seed system are generally taller than root-to-seed crops. Plants are generally taller at the higher plant densities but, as with carrots, closer spacing will increase the proportion of seeds in the primary umbels.

The effect of plant density, harvest date and method were investigated by Gray *et al.* (1985d), who reported that the proportion of primary umbel seed increased with plant density. They also found that maximum seed yield was obtained when the seed moisture content was between 50 and 70%, which occurred at approximately 46 days from flowering. Gray and Steckel (1985a) reported the effects of parsnip seed crop plant density, seed position on the mother plant, harvest date and method, and seed grading on embryo and seed size and seedling performance.

The plants are traditionally cut when the majority of the seeds in the primary umbel have ripened. This can be detected by their light-brown colour and by the splitting of the schizocarp to display two separate mericarps which are the 'seed'. Parsnip seed 'shatters' readily, and timing the cutting of the crop can be critical especially if there is the likelihood of inclement weather. As with carrots, some producers have used PVA glues to reduce loss by shattering.

Small areas of parsnip seed can be cut by hand, but larger areas are cut with a mower and left in windrows to dry. In some areas of North America, especially Oregon, the swather is taken over the field twice. The first cut is at about 75 cm high, which leaves the cut material perched on high stubble. This is followed by a second cut of the stubble under the swath, which brings all the material to ground level.

The seed can be threshed by a combine or stationary thresher, but as the dead or dry plant material (especially umbels) fracture easily, and will add to the plant debris which is later difficult to separate from the seed lot, the material is threshed when the crop is still damp from dew.

Final cleaning of the seed can be achieved by passing the material through aspirated screen separators.

Some workers are allergic to the sap of parsnip plants which can cause blisters on the exposed skin; it is therefore advisable to wear protective overalls, gloves and goggles when handling the crop and to seek early medical advice for workers who have any symptoms of an allergy.

Seed yield and 1000 grain weight

The average seed yield is about 1000 kg/ha, but some parsnip seed producers in Oregon, USA, can achieve double this.

The 1000 grain weight of parsnip is approximately 1 g.

Pathogens

The main seed-borne pathogens of parsnip are listed in Table 13.2.

Parsley: *Petroselinum crispum*

Origins and types

This crop is cultivated for its foliage which is used fresh as a green salad or dried as a herb, flavouring, garnish or fresh.

The wild species is a native of Europe, but cultivated forms are now spread throughout the world in temperate and tropical regions. There has been interest from time to time in the fresh leaves as a source of vitamin C, but most of the current commercial production is for dehydration.

Parsley is a biennial, but when grown as a commercial foliage crop it is treated as an annual. The seed crop is vernalized during the winter at the end of the first season.

There are two main types of leaf parsley, based on foliage characters, which are curled or plain. There are many selections of these two types, based on leaf

Table 13.2. The main seed-borne pathogens of parsnip; these pathogens may also be transmitted to the crop by other vectors.

Pathogens	Common names
Alternaria dauci	
Alternaria radicina	Black mould
Erysiphe heraclei	Powdery mildew
Itersonilia pastinacea	Canker
Phomopsis diachenii	
Pseudomonas viridiflava	
Strawberry Latent Ringspot Virus	

colour (intensity of chlorophyll), degree of curliness and petiole length. There is also a group of cultivars usually referred to as 'root parsley' or Hamburg parsley; cultivars in this group have a thickened tap root which may be cooked and eaten.

Cultivar description of parsley

Parsley cultivar descriptions are based on foliage characters (except for Hamburg parsley) and include the intensity of green colour, leaf crenulation and petiole length.

(A detailed test guideline (136) is obtainable from UPOV; see Appendix 2.)

Pollination and isolation

Parsley is predominantly cross-pollinated by a wide range of insects including honeybees and some genera of *Diptera*. Isolation should be at least 500 m between cultivars of the same leaf type, but increased to 1000 m between curled and smooth-leaved cultivars.

Agronomy

Parsley seed production requires similar soil types and nutrients as carrot. The seed takes up to 30 days to germinate under field conditions, therefore, a stale seedbed should be used especially where appropriate herbicides cannot be used, such as in organic crop production.

Seed-to-seed production
The seed is sown at a rate of 3 kg/ha in rows 56 cm apart during the late summer (July or August in the northern hemisphere).

Root-to-seed production
Seed is sown just before midsummer (May or June in the northern hemisphere). The higher sowing rate of up to 6 kg/ha is used for this method. Seed is sown in rows at about half the distance apart as compared with the seed-to-seed system, the exact distance depending upon the presence of irrigation furrows and the need for mechanical cultivations when residual herbicides are not used. An area of plants produced from the above sowing rate will provide sufficient transplants for up to 30 times the area when transplanted 25 cm apart in rows 90 cm apart. There is, however, a trend towards closer planting distances in order to reduce the timespan of seed maturing from successive umbels on individual plants.

Only the root-to-seed system should be used for production of all seed categories of Hamburg or root parsley, this ensures that the root branching tendency is kept to a minimum. The root-to-seed method should also be used for basic and other stock seed of the leaf types.

Roguing stages

1. Young plants in rosette stage before first autumn. Plants with annual habit (bolters in first year) must be removed. Check leaf type and characters, including petiole length and intensity of colour.
2. In lifted plants, root parsley cultivars, check that selected roots are relatively free of branching (fanged roots).
3. Second year, before bolting, check leaf characters, as for stage 1.

Seed harvesting

The ripe seed of parsley shatters easily, and it is therefore very important that the crop is cut at a stage to minimize loss, usually just before the primary umbels shatter.

The crop, which is between 1 and 2 m tall by the time that flowering finishes, is cut into windrows. Large-scale producers use a combine for separating the seed from the straw, but on a small scale the seed is separated by using stationary or mobile threshers.

Further separation and cleaning are later achieved by passing the seed through an aspirated screen cleaner.

Seed yield and 1000 grain weight

The average parsley seed yield is approximately 800 kg/ha, but it is possible to achieve higher yields than this at higher plant densities with a larger proportion of the seed ripening simultaneously.

The 1000 grain weight of parsley seed is 2 g.

Current trends

In the USA, some large-scale parsley seed producers take a second seed crop in the third year. This saves the cost of sowing and young plant management, although it can result in a high weed seed content if herbicides are not used.

An interesting innovation in the USA, especially in Oregon, is parsley for seed production being grown with an intercrop of dill (*Anthemum graveolens* L.). Dill is an annual crop, and the two species are sown together in the spring by a modified seed drill with two seed boxes. Seeds of the two species are sown in the same drills. The advantage of this intercropping is that seeds of the two species is

Table 13.3. The main seed-borne pathogens of parsley species; these pathogens may also be transmitted to the crop by other vectors.

Pathogens	Common names
Alternaria dauci	
Alternaria radicina	
Cercosporidium punctum	Leaf and stem rot
Erysiphe heraclei	Powdery mildew
Gibberella avenacea	Brown root rot
Phoma sp.	
Rhizoctonia petroselini	Root and basal stem rot
Septoria petroselini	Leaf spot
Parsley Latent Virus	
Strawberry Latent Ringspot Virus	

harvested in separate years. The dill's leaf canopy at the higher level is relatively thin and sufficient light reaches the parsley. The dill seed is cut and removed with its trash in the autumn of the first year but the parsley leaf rosettes are relatively undamaged. The parsley plants flower in the second year, and their seed is then harvested. This is a rare example of intercropping in mechanized vegetable seed production. The two species are resistant to the same herbicides, and producers maintain that energy, space and herbicides are saved.

Some large-scale parsley seed producers in the UK under sow the parsley with barley in the first year. This also provides an income during the year that the parsley is in the vegetative phase, but relies on high quality stock seed rather than on roguing for maintaining genetic quality.

Pathogens

The main seed-borne pathogens of parsley are listed in Table 13.3.

Celery: *Apium graveolens* L. var. *dulce* (Mill.) DC.

Origins and types

The modern cultivars of celery, which are popular vegetables used for salads and cooking in Europe and North America, have been developed from a wild marsh plant which is widely distributed in Europe and Asia.

There are two basic groups of celery cultivars: trench and self-blanching. The trench types are usually planted in shallow trenches and earthed-up ('blanched') during the latter part of their production for market. Some market growers tie paper collars around the celery plants to exclude the light and increase etiolation, but this practice is relatively uncommon now because of the high labour requirement and also because of very significant improvements by plant breeders of the self-blanching cultivars. The trench cultivars were originally developed for autumn and winter production, but the season has been considerably extended by the introduction of the self-blanching types that have gained very significantly in consumer popularity. There has been an increase in the production of heads or 'sticks' of these latter types as export crops in the USA, especially California, and in some Mediterranean countries, e.g. Israel and Spain. Self-blanching celery has also become an important protected crop for early production in northern Europe.

In some parts of the Middle East and Asia, the plants are cultivated at high density and the young leaves are cut, bunched and marketed locally for flavouring cooked dishes.

The petioles of self-blanching cultivars are succulent as a result of their wide, fleshy petioles, with a high ratio of parenchyma to vascular bundles. It is the strand of collenchyma inside the outer epidermis associated with each vascular bundle that is responsible for the 'stringiness' in celery petioles; the less 'stringy' cultivars having relatively small amounts of collenchyma. The self-blanching types have less frost resistance than the trench types.

Cultivar description of celery

- There are open-pollinated and hybrid cultivars.
- Type for production system: trench and self-blanching.
- Plant height, petiole characters, colour white, yellow green or pink; 'stringless' character, shape of transverse section of petiole.
- Leaf characters, colour, leaf tip colour.
- Resistance to bolting in first year.
- Resistant to specific pathogens, including *Septoria apiicola*.

(A detailed test guideline (082) is obtainable from UPOV; see Appendix 2.)

Soil pH, nutrition and irrigation

Celery requires a soil with a pH of between 6.5 and 7.5, and is not very tolerant of acid conditions. Soils with high organic matter contents are preferable.

The crop has a relatively high moisture requirement and should not be cultivated in areas of low summer rainfall unless there is an adequate water supply and a satisfactory irrigation system available. Irrigation is especially important in the crop's first year in order to produce heads of a satisfactory size.

The nutrient ratio required by the crop is 1:2:4 of N:P:K. Although the market crops (especially summer production of self-blanching types) respond to top dressings of nitrogen, these should be applied cautiously for seed production as they will increase susceptibility to frost damage. Some seed producers do, however, apply a top dressing of nitrogen in the spring of the second year.

Flowering

Celery is a biennial, but the market crops are cultivated as annuals. Generally, celery requires both short days and a low temperature for flower initiation, followed by long days to promote bolting and flowering. The physiology of celery, including vegetative development, bolting and flowering, has been reviewed by Pressman (1997).

After vernalization the plant grows to a height of 1 m in the second year, forming a very branched plant with a large number of compound umbels. The individual flowers are relatively small compared with other vegetable seed crops in *Apiaceae*.

Pollination and isolation

Celery flowers are self-fertile, but are largely cross-pollinated by insects. Minimum isolation distances are 500 m, although this should be increased to 1000 m for basic seed production and for isolation of different types.

When celery is flowering in polythene structures, access for pollinating insects is provided by leaving the entrance doorways open. However, if lack of sufficient isolation does not allow for this then blowflies are introduced to assist adequate pollination (Smith and Jackson, 1976).

Agronomy

The plants for seed production are either raised and planted out or are direct sown in the first year for seed production in the second year. If high genetic quality seed is required, it is necessary to sow at a similar date to that for the market crop. This allows for the seed crop to be grown in the appropriate environment during its early stages.

A later sowing time, after midsummer (similar to seed-to-seed technique for the root crops in *Apiaceae*) is used for the final multiplication stage to produce a commercial seed stock. This still provides the possibility of roguing plants for some vegetative characters such as petiole colour, leaf morphology and relative plant height, but it does not allow roguing out plants which bolt in the first year of a normal 2-year seed crop.

Another factor which must be taken into consideration when deciding on sowing and planting out times is the severity of winter in the proposed seed production area. There is an increasing trend in Europe to sow and plant out at times similar to the commercial fresh market crop and to plant selected plants in polythene tunnels before winter sets in.

For transplanted crops, the seed is usually sown under protection. There is an increased use of modules for plant raising. The young plants are hardened-off and planted out when they have developed about five to six leaves.

Plants are generally spaced in the field in rows 60 cm apart, with 30 cm between plants in the rows; these distances can be modified according to the irrigation system used, and whether or not the plants will be earthed up.

In areas with winters of continuous temperatures at or below freezing the plants must be protected. This is usually done by lifting the plants in the late autumn and storing them at temperatures just above freezing until replanting in the spring. Throughout the storage time the roots are kept moist and the foliage relatively dry. If stored in buildings or cellars, the relative humidity should be maintained at about 75%. Traditionally, cellar or pit storage has been used and the plants are transferred to the field in the spring. This operation is more labour-intensive and there is now an increasing use of polythene structures for growing on selected plants until seed is harvested.

Roguing stages

1. Planting out: check leaf and petiole characters, also plant vigour.
2. Vegetative stage, when crop close to normal fresh market maturity: remove early bolting plants and check foliage morphology, including leaf and petiole characters, colour, and length.
3. Lifted plants: check for absence of basal shoots, overall solidity of mature plant; discard early bolting plants.
4. Second year (before flowering): check general plant vigour and susceptibility to pathogens.

For basic seed production, special attention must be given to petiole characters, such as petiole ribbing, pithiness, and shape of transverse section

Seed harvesting

The seed crop is ready for harvesting when the plants have turned yellow, show signs of senility and the majority of seeds on the major inflorescences have become grey-brown colour.

Celery is notorious for shattering prior to seed harvest even in the best of conditions, and a lot of valuable seed can be lost if bad weather occurs or the cutting time is misjudged.

When produced in polythene structures or small areas outside, the plants can be pulled or carefully cut and placed on tarpaulins. In large-scale production, the crop is carefully cut, dried in windrows and combined before excessive shattering has occurred. If a combine is not used the material can be passed through a stationary thresher, but if dried on tarpaulins, especially under cover, the seed separates from the straw without machine threshing. The seed is further cleaned by an aspirated screen separator, but because the seed is very small, care must be taken in selection of the bottom screen.

Seed yield and 1000 grain weight

The average seed yield is about 500 kg/ha.

The 1000 grain weight is approximately 0.5 g.

Pathogens

The main seed-borne pathogens of celery and celariac are listed in Table 13.4.

Celeriac, Root Celery: *Apium graveolens* L. var. *rapaceum* (Mill.) DC

This crop is botanically very similar to celery, but the leaves, especially the petioles, are much less significant. The swollen stem bases and the upper part of the root form an edible storage organ, which is thick and similar in shape and size to a turnip, the swollen root is usually referred to as a 'tuber' which is of greater market importance than the leaves and is especially used as a winter salad, although the leaves are sometimes used on a garden scale.

The cultural requirements and other seed production techniques are the same as for celery, except that the plant spacing is less, usually 60 × 30 cm, giving a higher plant population per unit area.

The seed-borne pathogens are the same as for celery listed in Table 13.4.

Cultivar description of celeriac

- Season for market production.
- Use: fresh market, winter storage, processing.
- Foliage:
    - Vigour, intensity of green, leaf protection for 'root'.
    - Petiole, relative length, colour of bases.
    - Basal shoots: low number desirable, except for main growing point.
- Resistance to frost, early bolting and *Septoria apiciicola*.

(A detailed test guideline (074) is obtainable from UPOV; see Appendix 2.)

Table 13.4. The main seed-borne pathogens of celery and celeriac; these pathogens may also be transmitted to the crop by other vectors.

Pathogens	Common names
Alternaria dauci	
Alternaria radicina	Black mould, root rot
Botrytis cinerea	Grey mould
Gibberella avenacea	
Phoma apiicola	Celery root rot, black neck, scab, seedling canker
Septoria apiicola	Late blight, small leaf spot, large leaf spot
Verticilium albo-atrum	Wilt
Erwinia carotovora	Soft rot, crater spot
Pseudomonas apii	Bacterial blight
Strawbery Latent Ringspot Virus	

Roguing stages

1. Vegetative stage, in first season, check foliage morphology, vigour and pigmentation; discard early bolting plants.
2. 'Tuber', check relative size and height/width ratio, colour, root characters. For basic and stock seed production, check internal characters from a sample of tubers.

Further Reading

Carrot and other vegetables in *Apiaceae*

Rubatsky, V.E. and Quiros, C.F. (1999) *Carrots and Related Vegetable Umbelliferae.* CAB International, Wallingford, UK.
Shinohara, S. (1984) *Vegetable Seed Production Technology of Japan, Elucidated with Respective Variety Development Histories, Particulars*, Volume 1. Tokyo, Japan, pp. 269–304.

Parsley

Shinohara, S. (1984) *Vegetable Seed Production Technology of Japan, Elucidated with Respective Variety Development Histories, Particulars*, Volume 1. Tokyo, Japan, pp. 185–194.

14 *Alliaceae*

The only genus in this botanical family of importance to vegetable producers is *Allium* and, although there are a large number of cultivated species in the genus, the following are the most important.

Allium cepa L.	Onion, shallot
Allium porrum L. syn. *Allium ampeloprasum* L. var. *porrum*	Leek
Allium fistulosum L.	Japanese bunching onion, Welsh onion
Allium sativum L.	Garlic
Allium schoenoprasum L.	Chives
Allium tuberosum Rottl. ex Spreng	Chinese chives
Allium chinense G. Don.	Rakko

The two most important crops are onion and leek, which are normally grown from seed. Shallots (formerly considered *Allium ascalonicum* L. as a separate species from onion) are usually vegetatively propagated, although they can also be produced from seed. Japanese bunching onion is produced from seed or vegetative division. There are now hybrids between *Allium cepa* and *Allium fistulosum* that are cultivated as 'salad' or 'bunching' onions. Garlic and rakko are traditionally vegetatively propagated by bulbs ('cloves'); but Pooler and Simon (1994) have described a technique for the production of true seed from garlic. Chinese chives are generally multiplied by bulbils formed as a result of apomixis.

The physiology of onion and garlic has been reviewed by Brewster (1997); the scientific principles as related to the practice of the production of onions and other vegetable alliums, their biochemistry and food science have been described and discussed by Brewster (1994).

The medicinal values of the vegetable alliums in the diet have been recognized and they have increased in popularity (Block, 1985).

Onion: *Allium cepa* L.

Allium cepa probably originated from Afghanistan, Iran and Pakistan, but is now widely cultivated in many areas of the world including the tropics and the temperate regions (Astey *et al.*, 1982). The storage potential of the mature bulbs has led to onions becoming an important crop. The very young plants and developing and mature bulbs are used in a range of ways. There are several uses for onions in food processing including pickling, chutneys, sauces and dehydration.

As onion bulb formation is dependent on day length, there are specific day-length requirement groups for different latitudes, from types with a 16-h day-length requirement adapted to the northern and southern latitudes to 12-h day-length types suited to the tropics. This photoperiodic effect on bulbing was discussed by Jones and Mann (1963). More recently it has been demonstrated that temperature also has an important influence in that the rate of bulb formation increases with temperature increase. The physiology of growth, development and bulbing has been reviewed by Brewster (1994).

The range of cultivation areas of onion and its uses have led to a large number of cultivars and types. The horticultural divisions of onion include bunching onions (salad onions), dry bulb (including pickling onions) and shallots.

Cultivar description of onion

- Open-pollinated or F1 hybrid cultivar. Photoperiod for bulbing.
- Specific market use: salad (bunching), bulbs, processing (dry matter content and scale colour are of interest to dehydrators).
- Season: early, maincrop, or storage.
- Suitability for sowing at specific times of year, including overwintering.
- Foliage pose, colour and waxiness (see Fig. 14.1).

Fig. 14.1. Single individual rows of different bulbing onion cultivars displaying contrasting foliage and stem characters.

(Test Guideline number 046 for onion, echalion; shallot; grey shallot: test guideline number 161 for Japanese bunching onion and Welsh onion are obtainable from UPOV; see Appendix 2.)

Figure 14.1 illustrates single rows of onion cultivars showing a range of foliage and stem habits.

The methods used to describe onion responses to day length and temperature have been discussed by Currah and Astley (1996), who suggest that the conditions under which onion cultivar trials are conducted should be more exactly defined.

Soil pH and nutrition

Onions will grow satisfactorily on soils with a pH of 6.0–6.8.

The ratio of N:P:K applied during seedbed preparation is 1:2:2, although some bulb producers increase the nitrogen ratio according to the soil status. Very little work has been reported on the effects of nutrition in the first year on seed produced in the second year when grown as a biennial seed crop. Work by Ahmed (1982) showed that N, P and K applications equivalent to 150 kg/ha produced the largest bulbs and highest total bulb yield at the end of the first year, and that supplementary nitrogen application not exceeding 100 kg/ha in the second year applied during anthesis enhanced seed quality. Higher levels of nitrogen increased seed yield, but at the expense of seed quality. The higher potassium levels during bulb production were carried over to the second year and also enhanced seed quality.

Irrigation

Hawthorn (1951) found that high soil moisture in the seeding year favoured high seed yield. This was especially so at high plant densities with rows closer than 76 cm apart. Work by Millar et al. (1971) demonstrated that there is a large drop in water potential between the flowers and upper part of the seed stalk. They also reported lesser, but still important, differences between water potential of flowers and other parts of the plant. This work clearly demonstrated the moisture-sensitive stage of onions throughout anthesis. Borgo et al. (1993) reported work in Argentina that water stress from the start of bulb sprouting to the beginning of anthesis reduced the numbers of umbels and flowers per plant.

In practice, the soil surface should not be continuously wet because it will predispose the crop to infection by Botrytis allii (neck rot or damping off). Bulbs generally take longer to ripen in wet seasons or if unnecessary irrigation water is applied towards the end of the season.

Crop husbandry

The seed-to-seed and bulb-to-seed methods are both used in onion-seed production. The former does not allow for bulb inspection as relatively small plants overwinter from a summer sowing and the method is not suitable for basic seed production.

The bulb-to-seed method is a longer process, with mature bulbs produced by the end of the first season (as in the commercial production of bulbs for the fresh market). This allows roguing and selection to include mature bulb morphology and is especially important for basic seed production. The selected bulbs flower and seed in their second year.

Seed-to-seed

The seeds are drilled in midsummer to early autumn, depending on the cultivar and local climate. Generally the sowing date is towards the end of this period in warmer areas. The plants have to reach sufficient size for vernalization. Seed is sown at the rate of 4–5 kg/ha in rows 70–100 cm apart. Although the seed-to-seed crop cannot be examined for mature bulb characters the fields are rogued at the end of summer to remove obvious off-types, such as incorrectly coloured bulbs.

Bulb-to-seed

The seeds to produce bulbs for this method are sown either in single rows on the flat 40–45 cm apart or in beds with a distance of 90–100 cm between bed centres and rows within the beds 30–40 cm apart. A sowing rate of 3–6 kg/ha is higher than that adopted for the production of a commercial bulb crop for market, because a smaller bulb of approximately 5–8 cm diameter is preferred. One hectare of bulbs from the first year will plant 5 ha for the seed production. The bulbs selected for seed production are usually referred to as 'mother bulbs'.

When the tops die down at the end of the first growing season, the bulbs are lifted and dried. In some areas, the tops are trimmed off with a knife but Bleasedale and Thompson (1965) have shown that it is better for the tops to dry off naturally and to maintain aeration during this 'curing' period than to accelerate the process by trimming. Any necessary trimming is best done at the end of the curing period. The bulbs are rogued (or selected) at this time. They are then stored and examined again before replanting the following season. In some areas, the bulbs are replanted immediately after curing and sorting, but this can only be done where overwintering field conditions are likely to be suitable (either relatively mild or a covering of snow protects the overwintering bulbs). Mother bulbs are replanted in furrows 70–100 cm apart. A practical guide to the storage of bulbs in the tropics and the selection of suitable storage systems has been produced by Brice et al. (1997).

The agronomy of onion production in tropical regions (including important aspects of seed production) has been reviewed by Currah and Proctor (1990). Their findings, conclusions and recommendations were the results of conducting surveys in 46 tropical countries.

Flowering

There has been considerable progress in understanding the requirements for flowering of onions in recent years. Brewster (1987) and Rabinowitch (1990) have reviewed the research on onion flowering. The four broad stages in the life cycle of onion plants with particular reference to flowering have been described and discussed by Brewster (1994).

Pollination

The duration of anthesis is approximately 4 weeks on individual umbels, and there is a complicated succession of flowers opening on each. The sequence of flower opening and incidence of protandry in onion flowers has been studied by Currah and Ockendon (1978). The effects of environmental factors on onion pollen and ovule development were reviewed by Currah (1981) who also discussed pollination biology (Currah, 1990).

The flowers are pollinated by bees, flies and other insects. It is essential to ensure that there are sufficient populations of pollinating insects to achieve the full potential of onion seed, if in doubt supplementary beehives should be introduced into the fields. It is also possible in some situations to encourage the development of increased blowfly populations by distributing suitable carrion or dried fish among the flowering crop (Currah and Proctor, 1990).

Isolation

The minimum recommended isolation distance between different cultivars is 1000 m. Some seed-regulatory authorities allow shorter distances than this for cultivars with the same bulb colour. In some countries, there are declared zones in which only cultivars of a specific bulb colour can be grown for seed. Onion seed crops must also be isolated from any flowering multiplier types of onion and shallots.

Roguing stages

Seed-to-seed
1. During autumn of the first season: remove plants with off-type foliage, off-type bulb or stem colour and any plants bolting in their first year.
2. At the start of anthesis, in second year: remove plants with off-type foliage, off-type bulb or stem colour and check inflorescence characters where appropriate.

Bulb-to-seed
1. Before bulb maturity: remove plants with off-type foliage, off-type bulb or stem colour and any plants which are either late maturing or showing premature bolting.
2. When sorting lifted bulbs: check that morphology, colour and relative size of bulbs to be retained for seed production (mother bulbs) are true to type. Discard any bolting, bull-necked, bottle-shaped, split, doubles, damaged or diseased bulbs (Fig. 14.2).

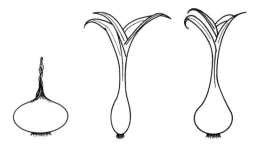

Fig. 14.2. Onion bulbs at the end of their first season: left, satisfactory bulb with thin neck; middle, 'bull-neck'; and right, a 'bottle-shaped' bulb.

3. At replanting (i.e. if the mother bulbs are not immediately replanted after previous roguing or selection at stage 2). Check characters as described above when sorting, also discard bulbs which are sprouting early.
4. At the start of flowering: when appropriate, check inflorescence and flower characters. Flowering heads of plants infected with *Ditylenchus dipsaci* ('bloat' or eelworm) tend to bend over; infected plants should be removed from the seed production site and burnt.

Production of F1 hybrid seed

Hybrid cultivars have been developed using cytoplasmic male sterility. The techniques used for the maintenance of parental lines have been described by Watts (1980). The usual ratio of male-to-female rows is 1:4 or 1:8. The pattern and ratio depend upon mechanization and also upon the amount of pollen produced by the female parent. In some cases where there is difficulty in synchronized flowering of the two parents, differential storage treatments are used for the bulb lots of the two parents. Either of the parent lines may require roguing according to the maintenance breeder's instructions to remove off-types (e.g. plants with male fertile flowers in a male sterile female parent).

Seeds produced on the male parents are either harvested or the rows are rotovated according to value, prior to harvesting the seed from the female parent. Care must also be taken to ensure that one or other parent is clearly marked at the ends of the rows as morphological differences are sometimes not obvious. Some seed producers paint the end few plants on each female row in order to reduce the risk of admixture when harvesting the seed crop.

The production F1 hybrid onion seed in polythene tunnels has been described by Dowker *et al.* (1985).

Basic seed production

Basic onion seed is produced only by the bulb-to-seed system. Mother bulbs are usually subjected to more critical criteria than for commercial seed. Bulb hardness is sometimes an additional criterion, although this is more appropriate for selecting material in breeding programmes (Fennel, 1978).

Harvesting

Traditionally onion seed heads have been harvested by hand when approximately 5% of the capsules on individual heads are shedding ripe seeds. The seeds are black when ripe and can be seen against the silvery coloured capsules. The seed heads shatter readily and the exact timing of this operation is based on experience and local weather conditions. Steiner and Akintobi (1986) have studied the effect of harvest maturity on subsequent seed viability; the variability of onion seed as influenced by temperature during seed growth has been investigated by Gray and Steckel (1984).

The seed heads with approximately 10–20 cm of scape (seed stalk) attached are removed by cutting with a sharp knife or secateurs. When cutting, the umbel is supported in the palm of the hand and held between the fingers to avoid loss of seed. In some areas, the crop is harvested with a once over cut, but in others successive harvests are made.

Although hand harvesting is still used in many onion seed production areas, there have been continued developments in the mechanization of harvesting. One system is to cut the stems at approximately 15 cm above ground level with mowers and pick up the cut material with an elevator which transfers it to mobile containers. The material is then further dried on sheets. The presence of the long stems in the heaps increases the incidence of the material heating up especially if the heaps are up to 200 cm deep, in which case they must be turned frequently, on at least a daily basis.

Work in Israel has shown that direct harvesting and threshing of the onion seed crop can lead to relatively high losses. As a result of this Globerson *et al.* (1981) examined the nature of flowering and seed maturation in relation to mechanical harvesting. They reported that the best time for mechanical harvesting was when the seeds had a dry-matter content of 60–70%; seeds which were dried while still in capsules attached to the stalks germinated better than those dried in their capsules after separation from the umbels.

After cutting, the seed heads are further dried on tarpaulins or sheets, either in the open or in suitable structures, the depth of cut umbels should not exceed approximately 20–30 cm, and the heads should be turned each day. Some producers use a tiered box or crate system to allow air to circulate freely.

When it is necessary to dry the material under cover, drying bins can be used, the draught temperature should not exceed 32 °C until the moisture content has been reduced to less than 19%, 38 °C until less than 10%, and 43 °C when below 10% (Brewster, 1994).

Threshing and cleaning

The material is ready for threshing as soon as it is dry, and the seeds can be separated from their capsules by rubbing in the hand. In order to avoid damaging the brittle seeds, threshing should not be delayed beyond this stage. Several threshing methods have been adopted for onions depending on the scale of operation and include flailing, rolling, threshing machines or combines. Onion seeds are very easily damaged during processing and frequent checks should be made to ensure that the seed coats are not accidentally cracked during any operation. Examination of samples with a hand lens will confirm if the processing is satisfactory. Concave settings should be adjusted to avoid injury to the seed. Work in Nepal, reported by Panthee and Subedi (1998), found that seed viability was greater after threshing by hand than sticks. Another important point is to ensure that the processing does not break too many of the former flower pedicels from their stalks as these are difficult to separate from the seed lot in subsequent cleaning processes. The light debris ('tailings') must also be examined occasionally to ensure that it does not contain an undue proportion of good seeds.

After threshing, the initial cleaning is usually achieved with an air-screen machine. Further upgrading according to the state of the seed lot can be done by passing the seeds over gravity tables. Seed lots with unacceptable levels of flower pedicels remaining can be further upgraded either by using a magnetic separator (the iron powder adheres to the pedicels and not the seeds) or by floatation. However, in the latter process it is a usual practice to put the light fraction only into water. During this process, which should not exceed 3 min per batch, the good quality seeds sink while the poor quality seeds and the pedicels float off, the heavier debris absorbs water and swells. After spin drying and further drying in racks the large debris is removed by an air-screen cleaner. The final onion seed lot must then be dried down to a moisture content not exceeding 12% or lower depending on the method of storage and packaging.

Pathogens

The main seed-borne pathogens of *Allium* are listed in Table 14.1.

Table 14.1. The main seed-borne pest and pathogens of *Allium* species; these pathogens may also be transmitted to the crop by other vectors.

Pathogens	Common names
Alternaria porri	Purple blotch
Botrytis allii	Damping-off, grey mould, neck rot
Botrytis byssoidea	Seedling damping-off, neck rot
Botrytis cinerea	Grey mould, collar rot, leaf spot
Cladosporium allii	Leaf blotch
Colletotrichum circinans	Smudge, damping-off
Fusarium spp.	
Perenospora destructor	Downy mildew
Pleospora herbarum	Black stalk rot, leaf mould
Puccinia allii	Rust
Sclerotium cepivorum	White rot
Urocystis cepulae	Smut
Onion mosaic virus	
Onion yellow dwarf virus	
Ditylenchus dipsaci	Bloat, eelworm rot, stem and bulb nematode

Seed yield and 1000 grain weight

The best yield from open-pollinated crops under ideal conditions is c.2000 kg/ha but c.1000 kg/ha is more common, and in some areas yields of c.500 kg/ha are more frequently accepted as satisfactory.

The yield from F1 hybrids is normally lower than from open-pollinated crops and is often as low as c.50–100 kg/ha.

The 1000 grain weight for onions is c.3.6 g.

Leek: *Allium porrum* L. syn. *Allium ampeloprasum* L. var. *porrum* and syn. *Allium ampeloprasum* L.

According to Astley *et al.* (1982) there are three cultivated forms in this species, i.e. leeks, kurrat and great-headed garlic (including pearl onions); Brewster (1994) lists these three but puts the pearl onion as an additional cultivated form.

Leeks have become an important vegetable crop in Europe, though especially as a winter crop in northern Europe, their marketing season has now been extended into spring, early summer and late summer. The technique of raising plants in modules has increased in recent years leading to a demand for high quality seed with a relatively high germination rate.

Kurrat is very similar to leek, but does not produce such a typical pseudo-stem, it is widely cultivated in the Middle East for its young leaves which are an important ingredient in cooked and raw salad dishes. It is usually cultivated as a smallholder or home garden crop at a high plant density in relatively small areas and harvested as a green bunched leafy vegetable. The crop is repeatedly cut from the same plot over a period of up to 2 years before a replacement crop is grown from seed.

Great-headed garlic ('elephant garlic') is propagated from cloves which are less pungent than garlic. It is mainly cultivated in the USA, southern Europe and Asia (especially India).

Within the species of *Allium ampeloprasum* and its synonyms, leeks are by far the most important crop type produced from seed. Currah (1986) has reviewed leek breeding.

Workers at Horticultural Research International (UK) have reported the development of F1 hybrid seed production by the use of a male parent which is stimulated to produce 300–400 bulbils from one plant (Anonymous, 1992).

Leek cultivars have been classified into eight types according to their morphological and horticultural characters (Brittain, 1988).

Cultivar description of leek

- Open-pollinated or hybrid cultivar.
- Maturity period, resistance to frost, resistance to early bolting.
- Height of shaft or column (i.e. the blanched petioles).
- Degree of bulbing at shaft base.

- Leaf characters (foliage above soil level is often referred to as the 'flag').
  - Leaf colour green, grey-green, tendency to become blue-green during periods of low temperature. Degree of waxiness.
  - Leaf shape, including relative length and width: the leaves of some cultivars have a characteristic 'keel' to the leaf; leaf attitude.

(A detailed test guideline (085) is obtainable from UPOV; see Appendix 2.)

Agronomy

The effects of seed crop plant density, transplant size, harvest date and seed grading were investigated by Gray and Steckel (1986a). They reported that large seed (2.00–2.24 mm) increased seedling emergence in the field by 13% compared with ungraded seed, and by 21% compared with the smallest seed (1.60–1.80 mm). In investigations into seed yields at a range of plant densities, it was found that for crops grown at 97 plants/m², seed yields were higher than the lower density of 11 plants/m². Over 3 successive years they found that at the higher density seed yield was increased 8.5-, 2.8- and 2.3-fold t/ha respectively (Gray and Steckel, 1986b).

Soil pH and nutrition

No experimental work relating to soil and nutrient requirements for leek seed production has been reported in the literature. However, the market crop is known to respond to soils with high organic matter content and dressings of up to 60 t/ha of bulky organic manures which are incorporated during site preparation in winter.

Leeks tolerate a slightly acid soil and a pH between 6.0 and 6.8 is satisfactory. Calcareous materials to increase the soil pH should not be applied at the same time as bulky organic manures. The ratio of N:P:K applied during the final stages of soil preparations is 2:2:1, but the nutrient values of any bulky organic manures which have been applied must also be taken into account when calculating the amounts of fertilizers to apply.

Plant production

Leeks are biennials and normal flowering occurs in the second year after vernalization. The plants are not stored during the winter but remain in the ground. Both seed-to-seed and root-to-seed systems are used for the production of basic seed.

Seed-to-seed method
The seeds are drilled in rows 30 cm apart at a rate of 2–3 kg/ha. At this sowing rate, there is no need for subsequent thinning if the seeds are sown evenly. If higher sowing rates are used, then it is necessary to thin the young plants as

soon as they are large enough to handle to approximately 10 cm apart in the rows. The seed-to-seed system can present weed-control problems and for this reason alone it is unpopular with some leek seed producers.

Root-to-seed method

This is the only method used for basic seed production and is often preferred for the production of commercial seed stocks although it involves extra labour costs for the transplanting operation.

Seeds are sown in early spring in seed beds at the rate of 4–5 kg/ha. After 8–12 weeks the seedlings are lifted and graded. Some producers trim the plants before grading. Plants which are approximately pencil thickness are replanted 8–10 cm apart in prepared drills which are 10 cm deep, with 70 cm between the rows. The transplants are either dibbled in or planted by machine.

Adamson (1952), working in Southern Vancouver Island, Canada, investigated time of sowing and transplanting: when field sowings were made by 15 May, there was no significant reduction in seed yield, although there was a tendency for the earliest sowings to produce more seeds. He found that a close spacing equivalent to c.5 cm within the rows for direct drilled crops gave a greater total seed yield per unit area than a spacing of c.15 cm, although the higher plant density gave a lower seed yield per plant. When Adamson compared the seed yield from a July transplanting at c.15 cm within the rows with a spring sowing that was thinned to the same distance within the rows, he found that the seed yields were similar and that there was no marked differences in seed germination, although he did not report on other aspects of seed quality; relatively late transplantings (i.e. in the Autumn) from seedbeds sown in early June resulted in low seed yields. It is generally recommended that a crop for seed should be sown at approximately the same time as a market crop and that late transplanting is avoided.

Roguing stages and main characters to be observed

Seed-to-seed method

1. During the summer of the first season: when the plants are well developed, check leaf colour, relative size and morphology, and also presence or absence of pigmentation according to cultivar.
2. At the start of anthesis: as above, and also check colour of inflorescences.

Root-to-seed method

1. When transplanting from seed beds: check leaf colour, relative size and morphology, and also presence or absence of pigmentation according to cultivar.
2. In late summer of first year: after transplants are established and actively growing, check as described for stage 1. Also remove any plants which are bolting in their first year and check that vigour of individual plants is according to the type.
3. During the second year, at the start of anthesis: as described above for stage 1, also check inflorescence colour.

Pollination and isolation

Pollination and isolation distances are similar to onion although there are normally no special zoning arrangements. Supplementary pollinating insects have to be introduced when the seed crop is produced in polythene tunnels or structures.

Basic seed production

The root-to-seed method described above is used. The plants are lifted and selected in the late autumn or early winter and selected according to the cultivar's morphological characters. Selected plants are immediately replanted individually with a trowel or dibber (see Fig. 14.3). In recent years, the trend has been to grow the selected plants in polythene tunnels or structures.

Harvesting, drying, seed extraction and processing

Leek seeds are harvested by hand as described for onions. Leeks tend to flower later than onions, and as their seed heads also ripen later, care must be taken not to lose seed as a result of inclement autumn weather. Seed heads may require artificial drying after harvesting. Adamson (1960) found that material which had been dried continuously at 26.7 °C resulted in seed with a germination

Fig. 14.3. Planting selected leeks during the autumn of their first year in a polythene tunnel for production of basic seed.

potential of at least 65%. Brewster (1994) states that the cut seed heads are dried at 20–35 °C for 4–6 weeks. Where controlled drying is not available, the cut heads can be dried on plastic sheets in a greenhouse or similar structure until the seeds separate from their capsules.

The plant density and harvest date effects have been examined by Gray et al. (1991), who estimated that harvests made at about 950 accumulated day degrees after 50% flowering (calculated with a base temperature of 6 °C and a ceiling of 270°) would give maximum seed yield and quality.

The extraction and processing of leek seed is the same as described for onion seeds.

Seed yield and 1000 grain weight

The yield is c.500 kg/ha although some producers achieve yields of up to 600 kg/ha. Bonnet (1976) reports yields of up to 900 kg/ha.

The 1000 grain weight is c.3.75 g.

Pathogens

The main seed-borne pathogens of *Allium* species are listed in Table 14.1.

Further Reading

Brewster, J.L. (1994) *Onions and Other Vegetable Alliums*, 1st edn. CAB International, Wallingford, UK.

15 *Gramineae*

This botanical family is of major agricultural importance. It includes cereal grain crops such as maize, rice, wheat, oats, millet and rye and also a wide range of fodder crops, e.g. prairie, tropical and temperate grasses. The important industrial crop sugarcane is also a member of *Gramineae*.

However, despite the very significant agricultural importance of this family, the only crops that are considered as vegetables are some vegetatively propagated bamboos of importance in Asia and the sweet corn group within the wide range of maize types (*Zea mays* L.).

Zea mays L.   Sweet corn, corn on the cob, vegetable maize

According to Purseglove (1972), the cultivars of *Zea mays* can be classified into seven groups.

1. Pod corn – *Zea mays tunica* Sturt.
2. Popcorn – *Zea mays everata* Sturt. (syn. *praecox*)
3. Flint maize – *Zea mays indurata* Sturt.
4. Dent maize – *Zea mays indenata* Sturt.
5. Soft or flour maize – *Zea mays amylacea* Sturt. (syn. *erythrolepis*)
6. Sweet corn – *Zea mays saccharata* Sturt. (syn. *sugosa* Bonof.)
7. Waxy maize – *Zea mays ceritina* Kulesh.

There are also ornamental forms, e.g. var. *japonica* Koern. with striped leaves and var. *gracillima* Koern., a dwarf form. The different cultivars of these groups cross-pollinate, and it is sometimes difficult to classify them exactly. The important vegetable group is sweet corn, and it contains the cultivars having a higher proportion of sugars to starches in the seeds' endosperm. It includes the supersweet cultivars with a genetically controlled factor for extra sugar in the seeds, resulting in a longer period for the change to starch after harvesting; they are, therefore, sweeter tasting at harvest and remain sweeter post-harvest. The several endosperm mutations of sweet corn have been described by Wolfe *et al.* (1997); they include sugary1 (su1), shrunken2 (sh2), sugary enhancer1 (se1), brittle (bt2), amylose extender (ae), dull (du) and waxy (w x). Hybrid cultivars of sweet corn are widely adopted in Europe and North America, although open-pollinated cultivars are still used in these areas and elsewhere.

Cultivar description of sweet corn

- There are open-pollinated and hybrid cultivars.
- Type of cultivar 'normal' or 'supersweet', genotype.
- Season of use and suitability for specific outlets (e.g. fresh market, canning or freezing).
- Plant height: relative height at start of anthesis.
- Time of start of anthesis: early, medium or late.
- Inflorescences: male ('tassel'), general colour of anthers; female ('silks'), time of emergence.
- Characters of cob (sometimes referred to as 'ears'): length and shape; number of rows of kernals (grain) in ear.
- Colour of grain at market maturity and when mature (dry).

(A detailed test guideline (002) is obtainable from UPOV; see Appendix 2.)

Soil pH and nutrition

Sweet corn tolerates a range of soil types, but it is important that the soil should have satisfactory water-retention characters. The crop tolerates a pH of 5.5–6.8, and appropriate quantities of lime should be applied if the pH is below this. The N:P:K fertilizer application during the final stages of soil preparation should be in the ratio of 1:2:2. Some growers apply a higher proportion of nitrogen in the base dressing, while others give a nitrogenous top dressing during the young plant stage.

Sowing and spacing

The sowing rate is regulated according to the relative seed size of the stock seed of the cultivar and seed lot. The larger-seeded cultivars are generally sown at a rate of up to 30 kg/ha and the smaller seeded types at 15 kg/ha in rows 70–100 cm apart depending on machinery used for cultivations and other operations. According to Wolfe *et al.* (1997) plant densities range from 74,000–86,000/ha.

Irrigation

As sweet corn is produced in areas with relatively low summer rainfall, it is important to ensure that sufficient irrigation is available. MacKay and Eaves (1962) found that sweet corn was very responsive to supplementary irrigation, especially from the pollen-shedding stage to cob maturity, and that removing the moisture stress by increasing irrigation increased the crop's response to nitrogen, phosphorus and potassium. The evidence for sweet corn response to water availability is generally in accord with the findings of workers with maize. Wolfe *et al.* (1988) found that water stress during anthesis adversely affects the synchrony of pollen release and the receptiveness of the 'silks'.

Flowering, pollination and isolation

The male flowers ('tassels') are in a terminal inflorescence and the female flowers ('silks') are borne in conjunction with the 'cobs' or 'ears' on lateral branches. There is a tendency for dehiscence of pollen up to 2–3 days in advance of the stigmas becoming receptive, which increases the incidence of cross-pollination.

The pollen is wind-borne and distances of up to 1 km may be required by some seed regulatory authorities to isolate seed crops of sweet corn from other types of *Z. mays*. The usual isolation distance is a minimum of 200 m. Some seed schemes allow a reduced isolation distance for hybrid seed production when the borders or perimeters of fields are planted with the male parent. Special attention should be given to the removal of any volunteer plants of *Z. mays* from seed production fields before the start of anthesis.

Roguing

Both parents must be inspected if hybrid seed is being produced, and roguing criteria applied according to the description of each parent line. The roguing operations must be completed before male and female inflorescences are fully developed.

An additional check for trueness to type (especially characters such as seed colour) can be made after harvesting when the husks are removed from the cobs.

Production of hybrid seed

Parental inbred lines for hybrid seed production are normally maintained by, or under the supervision of, the responsible maintenance breeder. Advice on ratio of female-to-male parent lines should be sought from the breeder in order to ensure adequate pollination and maximum yield of hybrid seed. The usual ratio of female-to-male lines is 4:2, although 4:1 may be adequate for the production of some hybrids. The relative height of the male parent can have some influence on this ratio because pollen distribution is more effective when the male parent is taller than the female.

Unless there is satisfactory genetic or chemical control of pollen production, the male inflorescences ('tassels') are removed from the female parent line. This operation, referred to as 'de-tasselling', must be done before the anthers dehisce; it may also be necessary to remove male inflorescences from the tillers of the female parent. Various methods for de-tasselling are used, including cutting by hand, working from mobile platforms or using machines with high level rotating cutters. It may be necessary to de-tassel a crop several times to ensure that self-pollination of the female parent does not take place. Whichever system is used it is important to ensure that the male inflorescences of the female parent are removed completely before they shed pollen.

Harvesting and processing

When hybrid seed is being produced, it is important to brief field workers on the layout and ratio of female-to-male lines before any harvesting commences. Clear instructions must be given to harvesting gangs regarding the disposal of mature cobs from the male lines.

The cobs of the seed crop are harvested when the seeds' moisture content is down to about 45% or less. Visual signs of this stage are the glazed appearance of the seeds as they harden. Wolfe *et al.* (1997), quoting Tallington of the Crookham Seed Company of Idaho, USA, stated that the percentage of moisture of the kernels at harvest is ideally about 35–40% for su1 genetic types, 40–45% for se1 genetic types, and 50% for sh2 types. Visual signs of the harvesting stage are the glazed appearance of the seeds as they harden. Small areas are harvested by hand but specialists in the USA either use mechanical pickers or mechanical combining. The husks are removed from the cobs immediately after harvesting leaving the 'ears' with the seeds attached. Further drying of the ears is either done in bird- and rodent-proof mesh cages known as 'cribs', or they are dried artificially. Wolfe *et al.* (1997) reported that the ears are dried at 35–38°C to the moisture content of 10–11%; they further state that the seed quality of sh2 types is significantly improved by a slow drying process and avoidance of temperatures above 38°C.

The operation of removing seeds from the cobs is usually referred to as 'shelling'. This is done with a maize or corn sheller (Fig. 15.1). Following the shelling operation the seeds may require separation from pieces of husk and other cob debris before being passed through an air-screen cleaner. The cleaned seed lot is usually graded in order to separate the different sized and shaped seeds originating from different parts of the cob.

Seed yield and 1000 grain weight

The seed yield depends on the cultivar, when hybrid seed is produced and the proportion of plants cultivated as the female parent. Yields of 1500 kg/ha are generally obtained, although specialist producers in the USA obtain up to 2500 kg/ha.

The 1000 grain weight of the super sweet cultivars is *c.*100 g and is *c.*210 g for 'normal' cultivars.

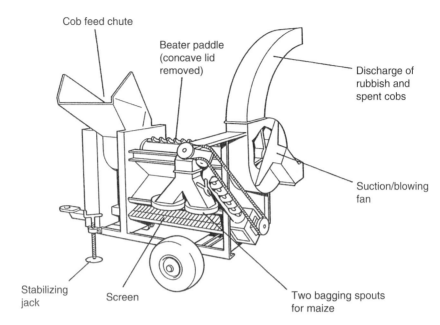

Fig. 15.1. Maize sheller. (Reproduced by courtesy of Alvan Blanch Ltd, Chelworth, Malmesbury, UK.)

Table 15.1. The main seed-borne pathogens of *Zea mays*; these pathogens may also be transmitted to the crop by other vectors.

Pathogens	Common names
Acremonium strictum	Kernal rot
Cochliobolus carbonum	Charred ear mould, southern leaf spot
Cochliobolus heterostrophus	Southern leaf spot, blight
Colletotrichum graminicola	Anthracnose
Diplodia frumenti	Dry ear rot, stalk rot, seedling blight
Fusarium spp.	
Gibberella fujikuroi	Gibberella ear rot, kernel rot, stalk rot
G. f. var. *subglutinans*	Seedling blight
G. zeae	Seedling blight, cob rot
Marasmius graminum	Seedling and foot rot
Penicillium spp.	Seed rot, blue-eye
Sclerophthora macrospore	Crazy top
Stenocarpella macrospora and *S. maydis*	Dry or white ear rot, stalk rot, seedling blight, root rot
Ustilaginoidea virens	False smut, green smut
Ustilago zeae	Smut, blister or loose-smut
Erwinia stewartii	Bacterial leaf blight, bacterial wilt, Stewart's disease, white bacteriosis
Maize leaf spot, Maize mosaic virus Sugar cane mosaic	

Pests and pathogens

Genetically modified lines with the ability to withstand losses from the Western corn borer and the European corn borer have been developed.

The main seed-borne pathogens of *Z. mays* are listed in Table 15.1.

Further Reading

Feistritzer, W.P. (ed.) (1982) *Technical Guideline for Maize Seed Technology*. FAO, Rome.

Wolfe, D.W., Azanza, F. and Juvik, J.A. (1997) Sweet corn. In: Wien, H.C. (ed.) *The Physiology of Vegetable Crops*. CAB International, Wallingford, UK, pp. 461–478.

16 *Amaranthaceae* and *Malvaceae*

Amaranthaceae

Amaranthus cruentus (L.) Sauer   African spinach, amaranth, Bush
greens, Chinese spinach

Amaranthus tricolor L
Amaranthus dubius C.
  Martius ex Thell.

All three of these *Amaranthus* species are known by many common names. According to Grubben (1977), these are the three *Amaranthus* species mainly cultivated as vegetables. *Amaranthus cruentus* is grown predominantly in Africa, *Amaranthus tricolor* is the main species grown in Asia, while *Amaranthus dubius* is cultivated in the Caribbean. Some of the *Amaranthus* species are grown as grain crops for the production of flour, especially in South America, the most important of which are *Amaranthus hypochondriacus* L., *Amaranthus caudatus* L. and to a lesser extent *Amaranthus cruentus*. A review of the grain amaranths has been made by Williams and Brenner (1995).

Cock's Comb, Green or White Soko, Quail Grass, Lagos Spinach: *Celosia argentia* L.

This species has been regarded as an underutilized crop by Badra (1993), who cites five other *Celosia* species as being of minor importance as a source of food, i.e. *C. trigyna, C. globerosa, C. insertii, C. leptostachya* and *C. pseudovirgata. Amaranthus* and *Celosia* are considered very important leafy vegetable species, especially for subsistence farmers and their dependants and are usually included in home garden schemes.

Cultivar description of *Amaranthus* and *Celosia*

- Genus of *Amaranthus* or *Celosia*.
- Species of *Amaranthus* or *Celosia*.
- Yield of leaf and/or grain per unit area and the percentage of dry matter content.
- General morphology and plant form. Apical dominance, degree of branching.
- Leaf: number from time of sowing or transplanting, shape and pigmentation.
- Inflorescence: time to flowering and specific day-length response colour and form.
- Resistance to specific pathogens, e.g. *Choanephora cucurbitarum* (Berk. et Rav.) *Thaxter* and *Meloidogyne* species.
- Resistance to drought conditions.

(A detailed test guideline (247) for Grain Amaranth is obtainable from UPOV; see Appendix 2.) Vernon (1999) has described some of the *Amaranthus* cultivars which have been derived from material originating in Central and South America.

Soil pH and nutrition

Although Grubben (1976) included liming as a variable in his investigations into the use of town refuse and NPK applications on amaranth production, there was no effect on yields from liming; however, it is possible that the quantities of lime applied were not sufficient to increase the soil pH. Other results of Grubben's investigations relating to pH indicate that these genera may exhibit lime-induced chlorosis and that while both crops tolerate a wide pH range, soils with a pH higher than 7.0 should not be used.

Grubben (1976) reported that the best response to fertilizers was obtained when the N:P:K ratios were 1:1:2. The response to potassium was especially noticeable and the plants readily exhibited potassium-deficiency symptoms on the poorer soils. Badra (1993) reported applications of up to 400 kg/ha of a similar N:P:K ratio for *Celosia argentea* as being beneficial.

Sowing and young plant management

The seed is usually broadcast in seed beds at the rate of $2 g/m^2$ for *Amaranthus* and $3 g/m^2$ for *Celosia*. These sowing rates are sufficient to produce 1000 seedlings ready for transplanting after approximately 3 weeks (Grubben, 1976). The seedlings are planted out in rows 60–70 cm apart with 40–50 cm between the plants. The sowing rate for both crops is 1 kg/ha when drilled direct.

It is normal practice to pinch out the growing point of those cultivars which have a relatively small apical inflorescence 4 weeks after planting to encourage the development and extension of secondary shoots. This is especially impor-tant with the cultivars which have strong apical dominance, but the growing points are not normally removed from cultivars which have a relatively large

Fig. 16.1. Sri Lankan seed stock growing in Malawi.

apical inflorescence. Figure 16.1 illustrates a tall cultivar from Sri Lanka in a 'growing out' trial in Malawi; note the strong apical dominance and large potential seed yield.

Flowering and pollination

There are neutral and short-day types in both genera which flower in all the tropical areas where the crop is cultivated.

Amaranthus is wind-pollinated and *Celosia* is predominantly insect-pollinated (Grubben, 1976).

Isolation

Plots for seed production should be isolated from all sources of pollen contamination by a minimum of 1000 m. There are wild forms of both species in the tropics which are common weeds. *Amaranthus spinosa* (the weed amaranth) can be a significant problem, and it is therefore important to ensure that the seed production plots and adjacent areas are kept weed-free. Off-types, rogues and a wild or weed *Amaranthus* are illustrated in Fig. 16.2.

Fig. 16.2. A mixture of desirable and undesirable plants from a seed crop of *Amaranthus*. From left to right: the local cultivar, two off-types resulting from crosses with a green leaf cultivar, a cross between the desired cultivar and the wild species, on the far right the thorny wild species.

Roguing stages

1. At planting out: check that plants are true to type, and that early flowering plants are removed; appropriate vegetative characters, including leaf number per plant, leaf shape and size according to plant stage are confirmed; the degree of leaf and stem pigmentation is confirmed according to cultivar.

2. Before start of anthesis: as above, in addition check that the plants are of appropriate height and that the degree of branching is according to the cultivar.

3. At start of anthesis: confirm appropriate plant height, time of flowering, inflorescence and individual flower colour, also general morphology of the plants.

Harvesting

The method of seed harvesting depends on the scale of production and to some extent on the cultivar. Seed crops of cultivars with apical dominance are cut when the seeds are mature, and there is a general yellowing of the plant. Seed heads are cut from the cultivars with axillary inflorescences as and when they mature; up to four successive harvests are made to obtain the maximum potential seed yield. Figure 16.3 illustrates a uniform *Amaranthus* seed crop with a very uniform stage of seed maturity, this crop will all be harvested in a single operation.

Fig. 16.3. A very uniform *Amaranthus* seed crop at Arusha, Tanzania, almost ready for a single harvest.

Seed yield and 1000 grain weight

Seed yields of up to 2000 kg/ha for *Amarantus* and 600 kg/ha for *Celosia* have been reported by Grubben (1976). The 1000 grain weight of *Amaranthus* is 0.3 g and for *Celosia* is 1 g.

Seed-borne pathogens

The main seed-borne pathogens of *Amaranthus* are *Alternaria amaranthi* (Peck) Van Hook (blight) and Strawberry Latent Ringspot Virus.

The main seed-borne pathogens of *Celosia* are Lilac Ring Mottle Virus and Spinach Latent Virus.

Malvaceae

Okra, Lady's Finger, Gombo: *Hibiscus esculentus* L. syn. *Abelmoschus esculentus* (L.) Moench.

Okra is an important vegetable in the tropics and subtropics. It probably originated in the Ethiopian region of Africa, but is now widely cultivated in Africa, especially in the Sudan, Egypt and Nigeria; it is also very important in other tropical areas including Asia, Central and South America. Although there has been an interest in growing it as a protected crop in heated greenhouses in northern Europe, it is mainly exported from the tropics to temperate areas.

Okra is cultivated for its immature edible fruits or 'pods'; the mature fruit are also dried and stored in some parts of Africa (e.g. Sudan) for local use in the high temperature season. The leaves are also used as a green vegetable in the tropics.

There are many local cultivars, many with very specific characteristics, e.g. the high mucilaginous content is preferred in the Sudan especially for storage as the dried crop. The modern cultivars produced in the USA, e.g. 'Clemson Spineless', fruit relatively early on more compact plants than the traditional African types.

Cultivar description of okra

- Use: fresh product, early production, processing, drying, export market, suitability for specific day length.
- Seed: colour of mature seed.
- Time of start of anthesis.
- Plant habit: degree of branching and side-shoot development.
- Stem: degree of anthocyanin pigmentation, length of internodes.
- Leaf: character of leaf from first leaf to fifth leaf, including lobing. Character of leaf from sixth leaf onwards, including lobing.
- Colour.
- Flower: relative size, colour of petals. Anthocyanin present or absent at base of petals, if present, on inner and/or outer surface.
- Fruit: relative length and shape. Shape in transverse section and number of facets.
- Colour, degree of spininess (pubescence).
- Resistance to specific pathogens, e.g. mosaic virus.

(A detailed test guideline (167) is obtainable from UPOV; see Appendix 2.)

Agronomy

Okra is grown as a warm-season crop throughout the tropics, and although the market crop can be produced in the rainy season, the seed crop is best timed so that pod ripening occurs under relatively dry conditions. The crop is grown as an annual, although there are a few types grown as perennials which originate mainly from Ghana. Seed ripening occurs approximately 1 month after the stage for normal harvesting of the fresh market crop.

Soil pH and nutrition

Okra is slightly tolerant of acid soil conditions and can be grown in soils with a pH between 6.0 and 6.8. The general fertilizer recommended is a N:P:K ratio of 1:2:1 applied during the final stages of site preparation. Nitrogenous fertilizers are also applied as top dressings at a rate of up to 50 kg/ha depending on the leaching rate early in the growing season.

Sowing

A sowing rate of 8 kg/ha is used for large-scale production of the more compact cultivars; sown on the flat in rows 45 cm apart. In areas where furrow irrigation systems are used, the sowing rate is 5–6 kg/ha with 90 cm between the rows. Plants are thinned to 15–30 cm apart in the rows according to the vigour of the cultivar. Some seed producers sow in blocks of 10–15 rows with a gap of two-row width between blocks to facilitate field inspections. The date of sowing and spacing has been investigated by Bisen *et al.* (1994) working in India; who reported that sowing in June with a plant spacing of 45 × 30 cm achieved the highest seed yields of over 1882 kg/ha.

Flowering

The majority of okra cultivars are short-day plants, requiring a minimum day temperature of 25 °C, but thriving in temperatures up to about 40 °C, although there are reports of exceptions to this. It is generally believed that the best seed crop is produced in areas with a relatively small day and night temperature differential.

In some tropical areas, the growing points are 'pinched' out or stopped in order to increase the flower number on the extended laterals, but this is not practised in large-scale production.

Pollination

The flowers are cross-pollinated by insects although self-fertilization may occur.

Isolation

Recommended isolation distances vary from one country's seed authority to another, but a minimum distance of 500 m is desirable.

Roguing stages and main characteristics to be observed

1. Before flowering: check the general plant height and habit; morphology of leaves; pigmentation of leaves, petioles and stems; remove plants with virus symptoms.
2. Flowering: check the relative size, pigmentation and colour intensity of flowers; remove plants with virus symptoms.
3. Fruiting: check that fruit are true to type; remove plants with virus symptoms.

Harvesting and seed extraction

There is a sequential ripening of okra pods on the plant. The pods of the angular-fruited types have a tendency to split when the seed ripens (see Fig. 16.4). The maturity of the seeds is associated with the pods becoming grey or brown according to the cultivar.

The traditional hand-harvesting of ripe pods is still done in many tropical areas where there is adequate labour, although the crop is combined when produced on a large scale in the USA.

Seeds are extracted after the hand-harvested pods become dry and brittle. The most efficient method of seed extraction by hand is to twist the pods open. Alternatively the pods are either flailed or the seeds extracted with a stationary thresher.

In some areas of the world, especially Malaysia, the initial pods are harvested as a fresh vegetable and the later pods are retained on the plants for seed production (Soo, 1977). This practice probably does not reduce the potential seed yield in areas with a sufficiently long growing season, as removal of the early pods tends to encourage further extension growth and flower development.

Fig. 16.4. Mature fruit of okra, showing sequential ripening from the bottom upwards, also splitting of the lowest fruit with potential seed loss before harvesting.

Table 16.1. The main seed-borne pathogens of *Hibiscus*; these pathogens may also be transmitted to the crop by other vectors.

Pathogens	Common names
Ascochyta abelmoschi	Ascochyta blight, pod spot
Botrytis sp.	Stem and capsule disease
Choanephora cucurbitarum	Okra fruit rot
Fusarium oxysporum f. sp. *vasinfectum*	
Fusarium solani	
Glomrella cingulata	
Rhizoctonia solani	
Okra leaf curl virus	
Mosaic virus	

Seed yield and 1000 grain weight

A relatively high seed yield of 1500 kg/ha is achieved in the okra-seed-producing areas of the USA but in many tropical countries the yield rarely exceeds 500 kg/ha.

The 1000 grain weight of okra seed is approximately 50 g.

Pathogens

The main seed-borne pathogens of *Hibiscus* are listed in Table 16.1.

Further Reading

Grubben, G.J.H. (1976) *The Cultivation of Amaranth as a Tropical Leaf Vegetable*. Royal Tropical Institute, Amsterdam.

Appendix 1: Vegetable Species that are Propagated Vegetatively

Crop	Scientific name	Method(s) of propagation (propagules)
Andean tubers		
Achira	*Canna edulis*	Division of rhizomes
Bitter potatoes	*Solanum × juzepczukii* and *S. × curtilobum*	Tubers Tubers
Oca	*Oxalis tuberose*	Selected tubers, cuttings of tuber shoots
Ulloco	*Ulluccus tuberosa*	Selected tubers and cuttings of tuber shoots
Mauka	*Mirabilis expansa*	Division of storage roots
Mashua, anu	*Tropaeolum tuberosum*	Selected tubers and cuttings of tuber shoots (also true seed)
Arrocacia	*Arrocacia xanthorrhiza*	Shoots produced on the crown
Artichoke, globe	*Cynarara cardunculus*	Side shoots with roots, un-rooted side shoots as cuttings
Artichoke, Jerusalem	*Helianthus tuberosus*	Selected tubers
Banana and plantains	*Musa*	Suckers, micro-corms, buds produced in humidity chamber or seed bed, tissue culture
Cassava	*Manihot esculenta*	Stem cuttings ('stakes')
Chinese artichoke	*Stachys affinis*	Tubers
Chinese chives, elephant garlic	*Allium tuberosum*	Division, bulbils formed by apomixis
Chives	*Allium schoenoprasum*	Division
Garlic	*Allium sativum*	Cloves (division of bulbs)
Ginger	*Zingiber officinale*	Tubers
Horseradish	*Armoracea rusticana*	Root cuttings

Continued

Crop	Scientific name	Method(s) of propagation (propagules)
Housa potato	*Solenostemen rotundifolius*	Tubers, stem cuttings, tissue culture
Japanese bunching onion, Welsh onion	*Allium fistulosum*	Division
Jicamba, yam bean	*Pachyrrizus erosus, P. tuberosus*	Tubers and tuber division
Konjac	*Amorphophallus konjac*	Corms, cormels
Livingstone potato	*Plectranthus asculentus,* syn. *Coleus esculentu*	
Madagascar or Sudan potato	*Coleus parviflorus*	Stem cuttings, selected small tubers
Potato (Irish potato)	*Solanum tuberosum*	Tubers, micropropagation
Rakkyo	*Allium chinense*	Bulbs
Skirret	*Sium sisarum* var. *sisarum*	Root cuttings
Sweet potato	*Ipomoea batatas*	Storage roots, vine cuttings
Tannia	*Xanthosoma sagittifolium*	Selected tubers, cuttings of tuber shoots
Taro	*Colocasia esculenta*	Selected corms, cuttings of corm shoots, runners (stolons)
Turmeric	*Curcuma domestica* and *C. xanthoriza*	Rhizomes
Yam	*Dioscorea* species	Selected tubers and cuttings of tuber shoots
Yam bean, Jicama, xiquima	*Pachyrrhizus erosus*	Selected tubers, stem cuttings from tubers
Water cress	*Rorippa nasturtium-aquaticum*	Stem cuttings
Water spinach, Kang Kong Green Engtsai	*Ipomoea aquatica*	Stem cuttings from selected plants
Welsh onion	*Allium fistulosum*	Division
Yacon, aricona	*Polymnia sonchifolia* syn. *P. edulis*	Cut pieces of tuber with eyes

Further Reading

FAO (2009) *Quality Declared Seed of Vegetatively Propagated Crops*. FAO, Rome.
Lebot, V. (2008) *Tropical Root and Tuber Crops: Cassava, Sweet Potato, Yams and Aroids*. CAB International, Wallingford, UK.
Onwueme, I.C. (1978) *The Tropical Tuber Crops*. Wiley, Chichester.

Appendix 2: UPOV Test Guidelines Relating to Vegetables

All test guidelines should be read in conjunction with UPOV document TG/1/3. The crop names and numbers are as listed by UPOV on their website: http://www.upov.int/en/publications/tg-rom-index.html

Title	UPOV Test Guideline Number
Asparagus	130
Beetroot	060
Bitter gourd	235
Black radish	063
Black salsify, *Scorzonera*	116
Broad bean	206
Brussels sprouts	054
Cabbage	048
Calabrese, sprouting broccoli	151
Carrot	049
Cauliflower	045
Celeriac	074
Celery, stalk celery/cutting celery, leaf celery, smallage	082
Chickpea	143
Chinese cabbage	105
Chinese chive	199
Chives	198
Cornsalad	075
Cucumber, gherkin	061
Cucurbita moschata	234
Curly kale	090
Dill	165

Continued

Title	UPOV Test Guideline Number
Eggplant	117
Endive	118
French bean	012
Garlic	162
Grain amaranth	247
Horse radish	191
Japanese bunching onion	161
Kohlrabi	065
Leaf beet, swiss chard	106
Leaf chicory	154
Leek	085
Lentil	210
Lettuce	013
Maize	022
Melon	104
Okra	167
Onion, echalion; shallot; grey shallot	046
Parsley	136
Parsnip	218
Peas	007
Potato	023
Pumpkin	155
Radish	064
Red cabbage	048
Rocket (garden rocket)	245
Rocket (wild rocket)	244
Runner bean	009
Rutabaga, swede	089
Savoy cabbage	048
Scorzonera	116
Shallot	046
Soybean	080
Spinach	055
Sprouting broccoli, calabrese	151
Squash	119
Swede, rutabaga	089
Sweet pepper, hot pepper, paprika, chilli	076
Swiss chard, leaf beet	106
Tomato	044
Turnip	037
Vegetable marrow, squash	119
Watermelon	142
Welsh onion, japanese bunching onion	161
White cabbage	048
Witloof chicory	273

References

Accatino, P. and Malagamba, P. (1982) *Potato Production from True Seed*. International Potato Center, Lima, Peru.

Adamson, R.M. (1952) Effects of various growing methods in leek seed production. *Scientific Agriculture* 32, 634–637.

Adamson, R.M. (1960) The effect of germination of drying leek seed heads at different temperatures. *Canadian Journal of Plant Science* 40, 666–671.

Ahmed, A.A. (1982) *The Influence of Mineral Nutrition on Seed Yield and Quality of Onion (Allium cepa L.)*. PhD Thesis, University of Bath, Bath, UK.

Andrews, J. (1984) *Peppers, the Domesticated Capsicums*. University of Texas Press, Austin, Texas, USA.

Anonymous (1980a) *International Code for the Nomenclature of Cultivated Plants*. The International Bureau for Plant Taxonomy and Nomenclature, Utrecht, pp. 12–13.

Anonymous (1980b) Top French seed starts from stumps. *Grower* 93(7), 16.

Anonymous (1992) *F1 Hybrid Leeks*. Annual Report of Horticulture Research International 1992–1993, pp. 62–63.

Anonymous (1994) *Genebank Standards*. Food and Agriculture Organization of the United Nations, Rome; International Board for Plant Genetic Resources, Rome.

Anonymous (1995) *Getting Farmers Involved in Research*. CGIAR News 2 No 2, pp. 9–10.

Anonymous (1996) *Technical Background Documents 6–11* World Food Summit, 5. The *Agricultural Research Agenda for the Next Decade*. FAO, Rome, p. 26.

Anonymous (2000) *Cartagena Protocol on Biosafety. The Secretariat of the Convention on Biodiversity*. World Trade Centre, Montreal, Canada.

AOSA (1983) *Seed Vigour Testing Handbook, Contribution No. 32 to the Handbook of Seed Testing*. Association of Official Seed Analysts, Ithaca, New York.

AOSA (1993) Association of Official Seed Analysts, Inc. Bylaws. *Newsletter of the Association of Official Seed Analysts*, 63(3), 95–105, Ithaca, New York.

AOSA (2000) *Vigor Testing Handbook*. AOSA, Ithaca, New York.

AOSA (2008) *Rules for Seed Testing*. AOSA, Ithaca, New York.

Astey, D., Innes, N.L. and van der Meer, Q.P. (1982) *Genetic Resources of Allium Species*. International Board for Plant Genetic Resources, Rome, 8 pp.

Astey, D. (2007) Banking on Genes. *The Horticulturist* 16(2), 2–5.

Atherton, J.G., Basher, E.A. and Brewster, J.L. (1984) The effects of photoperiod on flowering in carrot. *Journal of Horticultural Science* 59, 213–215.

Atherton, J.G. (ed.) (1987) *The Manipulation of Flowering*. Butterworths, London.

Atherton, J.G., Craigon, J. and Basher, E.A. (1990) Flowering and bolting in carrot: I. Juvenility; cardinal temperatures and thermal times for vernalization. *Journal of Horticultural Science* 65, 423–429.

Atkins, E.L., Anderson, L.D., Kellum, D. and Neuman, K.W. (1977) *Protecting Honey Bees from Pesticides*. Leaflet 2883, Division of Agricultural Sciences, University of California.

AVRDC (1998) *Vegetables for Poverty Alleviation and Healthy Diets*. AVRDC, Taiwan 29 pp.

Austin, R.B. (1972) Effects of environment before harvesting on viability. In: Roberts, E.H. (ed.) *Viability of Seeds*. Chapman and Hall, London and Syracuse University Press, Syracuse, New York, pp. 114–149.

Austin, R.B. and Longden, P.C. (1966) The effects of manurial treatments on the yield and quality of carrot seed. *Journal of Horticultural Science* 41, 361–370.

Badra, T. (1993) Lagos spinach (Celosia sp.). In: Williams, J.T.(ed.) *Underutilized Crops: Pulses and Vegetables*. Chapman and Hall, London, pp. 131–163.

Baes, P. and Van Cutsem, P. (1992) Chicory seed lot variety identification by leucine aminopeptidase and esterase zymogram analysis. *Electrophoresis* 20, 885–886.

Baker, K.F. (1972) Seed pathology. In: Kozlowski, T.T. (ed.) *Seed Biology*. Academic Press, New York and London.

Banga, O. (1963) *Main Types of the Western Carotene Carrot and their Origin*. W.E.J. Tjeenk Willink, Zwolle.

Banga, O. (1964) Identification of Western orange carrot varieties (Daucus carota L.). *Proceedings of the International Seed Testing* Association 29(4), 957–961.

Banga, O. (1976) Radish. In: Simmonds, N.W. (ed.) *Evolution of Crop Plants*. Longman, London and New York.

Barbuzzi, G. and Silviero, P. (1998) The production of seed. *Supplento* 54(24), 29–30.

Barre, H.J. (1963) Important equipment for drying seeds in North America. *Proceedings*

of the International Seed Testing Association 28(4), 815–826.

Bateman, A.J. (1946) Genetical aspects of seed-growing. *Nature* 157, 752–755.

Bay, A.P.M., Taylor, A.G. and Bourne, M.C. (1995) The influence of water activity on three genotypes of snap beans in relation to mechanical damage. *Seed Science and Technology* 23, 583–593.

Bedford, L.V. (1975) A comparison of single row and larger plot techniques for variety performance of carrots. *Journal of the National Institute of Agricultural Botany* 13, 349–354.

Benjamin, L.R., McGarry, A. and Gray, D. (1997) The root vegetables: beet, carrot, parsnip and turnip. In: Wien, H.C. (ed.) *The Physiology of Vegetable Crops*. CAB International, Wallingford, UK, pp. 553–580.

Biddle, A.J. (1981) Harvesting damage in pea seed and its influence on vigour. *Acta Horticulturae* 111, 243–247.

Biddle, A.J. and King, J.M. (1977) Effect of harvesting on pea seed quality. *Acta Horticulturae* 83, 77–81.

Bisen, R.K., Surendra, J. and Bisen, C.S. (1994) Effect of sowing dates and spacing on the yield of okra seed (*Abelmoschus esculentus* (L.) Moench). *Advances in Plant Sciences* 7(2), 244–250.

Blackwall, F.L.C. (1961) Factors affecting the set of runner bean pods, *Annual Report of The National Vegetable Research Station*, 1960, 40.

Blackwall, F.L.C. (1962) Factors affecting the set of runner bean pods, *Annual Report of The National Vegetable Research Station*, 1961, 48–49.

Blackwall, F.L.C. (1963) Factors affecting the set of runner bean pods, *Annual Report of The National Vegetable Research Station*, 1962, 49.

Bleasdale, J.K.A. (1964) The flowering and growth of watercress (*Nasturtium officinale* R. Br.). *Journal of Horticultural Science* 39, 277–283.

Bleasdale, J.K.A. and Thompson, R. (1965) Onion skin colour and keeping quality. *Report of the National Vegetable Research Station* 16, 47–49.

Block, E. (1985) The chemistry of onions and garlic. *Scientific American* 252, 94–99.

Bohart, G.E. and Nye, W.P. (1960) *Insect Pollinators of Carrots in Utah.* Bulletin 419, Agriculture Experiment Station, Utah State University, pp. 1–16.

Bohart, G.E., Nye, W.P. and Hawthorn, L.R. (1970) *Onion pollination as affected by different levels of pollinator activity.* Bulletin 482, Agriculture Experiment Station, Utah State University.

Bonina, J. and Cantliffe, D.J. (2005) *Seed Production and Seed Sources of Organic Vegetables.* IFAS Extension, University of Florida.

Bonnet, B. (1976) The leek (*Allium porrum* L.): botanical and agronomic aspects. Review of literature. *Saussurea* 7, 121–155.

Borgo, R., Stahlschmidt, D.M. and Tizio, R.M. (1993) Preliminary study on water requirements in onion (*Allium cepa* L.) cv. Valcatorce in relation to seed production. *AgriScientia* 10, 3–9.

Bowring, J.D.C. (1969) The identification of varieties of lettuce (*Lactuca sativa* L.). *Journal of the National Institute of Agricultural Botany* 11, 499–520.

Bowring, J.D.C. (1970) The identification of varieties of runner bean (*Phaseolus coccineus* L.). *Journal of the National Institute of Agricultural Botany* 13, 168–185.

Bowring, J.D.C. (1998) Organisation for economic co-operation and development seed schemes. In: Kelly, A.F. and George, R.A.T. (eds) *Encyclopaedia of Seed Production of World Crops.* Wiley, Chichester, UK, pp. 23–26.

Bradnock, W.T. (1998a) Association of Official Seed Certifying Agencies (AOSCA). In: Kelly, A.F. and George, R.A.T. (eds) *Encyclopaedia of Seed Production of World Crops.* Wiley, Chichester, UK, pp. 28–31.

Bradnock, W.T. (1998b) Association of Official Seed Analysts (AOSA). In: Kelly, A.F. and George, R.A.T. (eds) *Encyclopaedia of Seed Production of World Crops.* Wiley, Chichester, UK, pp. 34–36.

Brewster, J.L. (1987) Vernalization in the onion – a quantitative approach. In Atherton, J.G. (ed.) *The Manipulation of Flowering.* Butterworths, London, pp. 171–183.

Brewster, J.L. (1994) *Onions and Other Vegetable Alliums,* 1st edn. CAB International, Wallingford, UK, 236 pp.

Brewster, J.L. (1997) Onions and garlic. In: Wien, H.C. (ed.) *The Physiology of Vegetable Crops.* CAB International, Wallingford, UK, pp. 581–619.

Brice, J., Currah, L., Malins, A. and Bancroft, R. (1997) *Onion Storage in The Tropics: A Practical Guide to Methods of Storage and Their Selection.* Natural Resources Institute, Chatham, UK.

Brittain, M. (1988) Leeks-the long and the short of it. *The Grower* 110(3), 20–23.

Broadbent, L. (1976) Epidemiology and control of tomato mosaic virus. *Annual Review of Phytopathology* 14, 75–96.

Brouwer, W. (1949) Steigerung der Erträge der Hülsenfrüchte durch Beregnung sowie Fragen der Bodenuntersuchung und Düngung, *Zeitschrift fur Acker und Pflanzenbau* 91, 319–346.

Brouwer, W. (1959) *Die Feldberegnung,* 4th edn. DLG-Verlag, Frankfurt/Main.

Browning, T. and George, R.A.T. (1981a) The effects of nitrogen and phosphorus on seed yield and composition in peas. *Plant and Soil* 61, 485–488.

Browning, T. and George, R.A.T. (1981b) The effects of mother plant nitrogen and phosphorus nutrition on hollow heart and blanching of pea (*Pisum sativum* L.) seed. *Journal of Experimental Botany* 32 (130), 1085–1090.

Browning, T., Gavras, M. and George, R.A.T. (1982) *Proceedings XXI International Horticultural Congress,* Hamburg.

Caetano-Anolles, G. (1996) Fingerprinting nucleic acids with arbitrary oligonucleotide primers. *Agro-Food Industry Hi-Tech,* Jan/Feb, 26–35.

Carter, T.E. Jr. and Shanmugasundarum, S. (1993) Vegetable soybean (Glycine). In: Williams, J.T. (ed.) *Underutilized Crops: Pulses and Vegetables.* Chapman and Hall, London, pp. 219–239.

Chopra, K.R. and Chopra, R. (1998) Contract seed production and the role of seed growers. In: Kelly, A.F. and George, R.A.T. (eds) *Encyclopaedia of Seed Production of World Crops.* Wiley, Chichester, UK, pp. 9–14.

Chotiyarnwong, A., Chotiyarnwong, P., Gong-in, W., Nalampang, A., Potan, V., Benjasil, V. and Kajornmalle, V. (1996) Chiang Mai 1 – a vegetable soybean released in Thailand. *Tropical Vegetable Information Service Newsletter* 1(2), 12.

Choudhury, B. (1966) Exploiting hybrid vigour in vegetables. *Indian Horticulture* 10, 56–58.

Chowings, J.W. (1974) Vegetable variety performance trials technique – brassica crops. *Journal of the National Institute of Agricultural Botany* 13, 168–185.

Chowings, J.W. (1975) Vegetable variety performance trials technique celery, lettuce, leeks, onions and sweetcorn. *Journal of the National Institute of Agricultural Botany* 13, 355–366.

Chroboczek, E. (1934) A study of some ecological factors influencing seed-stalk development in beets (*Beta vulgaris* L.). *Cornell University Agricultural Experimental Station Memo* 154, 1–84.

CIP (1979) *Report of the Planning Conference on the Production of Potatoes from True Seed.* International Potato Center, Lima Peru, 172 pp.

Clarkson, C. and George, R.A.T. (1985) A comparison of identification characters for lettuce seedlings (*Lactuca sativa* L.). *Journal of the National Institute of Agricultural Botany* 17, 129–138.

Cochran, H.L. (1943) Effect of stage of fruit maturity at time of harvest and method of drying on the germination of pimento seed. *Proceedings of the American Society for Horticultural Science* 43, 229–234.

Cooke, G.W. (1982) *Fertilizing for Maximum Yield*, 3rd edn. Granada Publishing, London.

Cooke, R.J. (1992) *Electrophoresis Testing.* International Seed Testing Association, Zurich, 36 pp.

Cooke, R.J. (1995a) Gel electrophoresis for identification of plant varieties. *Journal of Chromatography* 698, 281–199.

Cooke, R.J. (1995b) Varietal identification of crop plants. In: Skerritt, J. and Appels, R. (eds) *New Diagnostics in Crop Sciences.* CAB International, Wallingford, UK, pp. 33–63.

Cooke, R.J. (1995c) Variety identification: modern techniques and applications. In: Basra, A.S. (ed.) *Seed Quality: Basic Mechanisms and Agricultural Implications.* Food Products Press, New York, pp. 279–318.

Cooke, R.J. and Reeves, J.C. (1998) Cultivar identification: review of new methods. In: Kelly, A.F. and George, R.A.T. (eds) *Encyclopaedia of Seed Production of World Crops.* Wiley, Chichester, UK, pp. 82–102.

Cooke, R.J., Higgins, J., Morgan, A.G. and Evans, J.L. (1985) The use of a vanillin test for the detection of tannins in cultivars of *Vicia faba* L. *Journal of the National Institute of Agricultural Botany* 17, 139–143.

Cooper, W. and MacLeod, J. (1998) Introduction to genetically modified crops. *Plant Varieties and Seeds* 11, 131–141.

Copeland, L.O. and McDonald, M.B. (1995) *Principles of Seed Science and Technology*, 3rd edn. Chapman & Hall, New York, 409 pp.

Craigon, J., Atherton, J.G. and Basher, E.A. (1990) Flowering and bolting in carrot: II. Prediction in growth room, glasshouse and field environments. *Journal of Horticultural Science* 65 (5), 547–554.

Crisp, P. and Gray, A.R. (1979) Successful selection for curd quality in cauliflower, using tissue culture. *Horticultural Research* 19, 49–53.

Crisp, P. and Lethwaite, J.J. (1974) Curd grafting as an aid to cauliflower breeding. *Euphytica* 23, 114–120.

Crisp, P. and Walkey, D.G.A. (1974) The use of asceptic meristem culture in cauliflower breeding. *Euphytica* 23, 305–313.

Crisp, P., Gray, A.R. and Jewell, P.A. (1975a) Selection against bracting defect in cauliflower. *Euphytica* 24, 459–465.

Crisp P., Jewell, P.A. and Gray, A.R. (1975b) Improved selection against the purple colour defect in cauliflower. *Euphytica* 24, 177–180.

Cromarty, A.S., Ellis R.H. and Roberts, E.H. (1982) *The Design of Seed Storage Facilities for Genetic Conservation.* International Board for Plant Genetic Resources, Rome.

Currah, L. (1981) Onion flowering and seed producton. *Scientific Horticulture* 32, 26–46.

Currah, L. (1986) Leek breeding: a review. *Journal of Horticultural Science* 61, 407–415.

Currah, L. (1990) Pollination biology. In: Rabinowitch, H.D. and Brewster, J.L. (eds) *Onions and Allied Crops*, Vol. 1. CRC Press, Boca Raton, Florida, pp. 135–149.

Currah, L. and Astley, D. (1996) Describing onions: can the definition of day-length descriptors be improved? In: Currah, L. and Orchard, J. (eds) *Onion Newsletter for the Tropics* 7, Natural Resources Institute, London, pp. 77–81.

Currah, L. and Msika, L. (1994) Cheap methods of storing small quantities of onion seeds without refigeration. In: Currah, L. and Orchard, J. (eds) *Onion Newsletter for the Tropics* 6, Natural Resources Institute, London, pp. 62–62.

Currah, L. and Ockenden, D.J. (1978) Protandry and the sequence of flower opening in the onion (Allium cepa L.). *New Phytologist* 81, 419–428.

Currah, L. and Proctor, F.J. (1990) *Onions in Tropical Regions*. Natural Resources Institute Bulletin No. 35. Natural Resources Institute, London, 232pp.

Dark, S.O.S. (1971) Experiments on the cross-pollination of sugar beet in the field. *Journal of the National Institute of Agricultural Botany* 12, 242–266.

Davis, J.H.C. (1997) Phaseolus Beans. In: Wien, H.C. (ed.) *The Physiology of Vegetable Crops*. CAB International, Wallingford and New York, pp. 409–428.

Dearman, J., Brocklehurst, P.A. and Drew, R.L.K. (1986) Effects of osmotic priming and ageing on onion seed germination. *Annals of Applied Biology* 108, 639–648.

DEFRA (2003) *The Economic and Agronomic Feasibility of Organic Seed Production in the United Kingdom, and its Subsequent Quality*. DEFRA, London.

Delouche, J.C. (1980) Environmental effects on seed development and seed quality. *Hortscience* 15, 13–18.

Doku, E.V. (1969) Growth habit and production in bambara groundnut (Voandzeia Subterranean). *Ghana Journal of Agricultural Science* 2, 91–95.

Douglas, J.E. (1980) *Successful Seed Programmes*. Westview Press, Boulder, Colorado, 302 pp.

Dowker, B.D., Bowman, A.R.A. and Faulkner, G.J. (1971) The effect of selection during multiplication on the bolting resistance and internal quality of Avon early red beet. *Journal of Horticultural Science* 46(3), 307–311.

Dowker, B.D., Currah, L., Horabin, J.F., Jackson, J.C. and Faulkner, G.J. (1985) Seed production of an F1 hybrid onion in polythene tunnels. *Journal of Horticultural Science* 60, 251–256.

Draper, S.R. and Keefe, P.D. (1989) Machine vision for the characterisation and identification of cultivars. *Plant Varieties and Seeds* 2, 53–62.

Drew, R.L.K., Hands, L.J. and Gray, D. (1997) Relating the effects of priming to germination of unprimed seeds. *Seed Science and Technology* 25(3), 537–548.

Drijfhout, E. (1981) Maintenance breeding of beans. In: Feistritzer, W.P. (ed.) *Seeds, FAO Plant Production and Protection Paper 39*. FAO, Rome, pp. 140–150.

Eguchi, T., Matsomura, T. and Ashizawa, M. (1958) The effect of nutrition on flower formation in vegetable crops. *Journal of American Horticultural Science* 72, 343–352.

Eisa, H.M. and Wallace, D.H. (1969) Morphological and anatomical aspects of petaloidy in carrot. *Journal of The American Society for Horticultural Science* 94, 545–548.

Ellis, R.H. and Roberts, E. (1981) The quantification of ageing and survival in orthodox seeds. *Seed Science and Technology* 9, 373–409.

Ellis, R.H., Hong, T.D., Astley, D., Pinnegar, A.E. and Kraak, H.L. (1996) Survival of dry and ultra-dry seeds of carrot, groundnut, lettuce, oilseed rape and onion during five years' hermatic storage at two temperatures. *Seed Science and Technology* 24(2), 347–358.

Eponou, T. (1996) Linkages between research and technology users in Africa: the situation and how to improve it. ISNAR Briefing

Paper Number 31, International Service for National Agricultural Research, The Hague, 8 pp.

FAO (1993) *Quality Declared Seed*. FAO Plant Production and Protection Paper 117, FAO, Rome, 186pp.

FAO (1995) *Proceedings of Regional Workshop on Improved On-farm Seed Production for SADC Countries*. FAO, Rome, 43pp.

FAO (2004). *First World Conference on Organic Seed*. The Proceedings are published jointly by The International Federation for Organic Movements (IFOAM), The International Seed Federation ISF) and FAO, Rome.

FAO (2006) *Quality Declared Seed System*. FAO Plant Production and Protection Paper, FAO, Rome, 243 pp.

Faulkner, G.J. (1962) Blowflies as pollinators of brassica crops. *Commercial Grower* 3457, 807–809.

Faulkner, G.J. (1971) The behaviour of honey-bees (*Apis mellifera*) on flowering Brussels sprout inbreds in the production of F1 hybrid seed. *Horticultual Research* 11, 60–62.

Faulkner, G.J. (1978) Seed production of F1 hybrid Brussels sprouts. *Acta Horticulturae* 83, 37–42.

Feistritzer, W.P. (1981) The FAO Seed Improvement and Development Programme (SIDP). *Seed Science and Technology* 9(1), 37–45.

Fennell, J.F.M. (1978) Use of a durometer to assess onion bulb hardness. *Experimental Agriculture* 14, 169–172.

Fennell, J.F.M. and Dowker, B.D. (1979) Screening cultivars for characteristics required in breeding autumn-sown onions. *Journal of the National Institute of Agricultural Botany* 15, 104–112.

Fitzgerald, D.M., Barry, D., Dawson, P.R. and Cassels, A.C. (1997) The application of image analysis in determining sib proportion and aberrant characterization in F1 hybrid Brassica populations. *Seed Science and Technology* 25(3), 503–509.

Forsberg, G., Anderson, S. and Johnsson, L. (2000), Evaluation of hot, humid air seed treatment in thin layers and fluidized beds for seed pathogen sanitation. *Journal of Plant Disease and Protection* 109, 357–370.

Frankel, R. and Galun, E. (1977) *Pollination Mechanisms, Reproduction and Plant Breeding*. Springer, Berlin, Heidelberg and New York, 281 pp.

Free, J.B. (1970) *Insect Pollination of Crops*. Academic Press, London, 544 pp.

Free, J.B. and Racey, P.A. (1968) The pollination of runner beans (Phaseolus multiflorus) in a glasshouse. *Journal of Apicultural Research* 7(2), 67–69.

Garry, G., Jeuffroy, M.H. and Tivoli, R. (1998) Effects of ascochyta blight *Mycosphaerella pinodes* (Berk. and Blox.) on biomass production, seed number and seed weight of dried-pea (*Pisum sativum* L.) as affected by plant growth stage and disease intensity. *Annals of Applied Biology* 132 (1), 49–59.

Gaunt, R.E. and Liew, R.S.S. (1981) Control of diseases in New Zealand broad bean seed crops. *Acta Horticulturae* 111, 109–112.

Gavras, M. (1981) *The Influence of Mineral Nutrition, Stage of Harvest and Flower Position on Seed Yield and Quality of Phaseolus vulgaris L.* PhD Thesis, University of Bath, England.

George, R.A.T. (1978) The problems of seed production in developing countries. *Acta Horticulturae* 83, 23–29.

George, R.A.T. and Evans, D.R. (1981) A classification of winter radish cultivars. *Euphytica* 30, 483–492.

George, R.A.T., Stephens, R.J. and Varis, S. (1980) The effect of mineral nutrients on the yield and quality of seeds in tomato. In: Hebblethwaite, P.D. (ed.) *Seed Production*. Butterworths, London and Boston, pp. 561–567.

Globerson, D., Sharir, A. and Eliasi, R. (1981) The nature of flowering and seed maturation of onions as a basis for mechanical harvesting of the seeds. *Acta Horticulturae* 111, 99–114.

Gormley, T.R. (1989) Studies on the role of fruit and vegetables in the diet. *Professional Horticulture* 3, 66–68.

Goulson, B., Lye, D.G. and Darvill, B. (2008) Decline and conservation of bumble bees. *Annual Review of Entomology* 53, 191–208.

Gray, D. (1979) The germination response to temperature of carrot seeds from different umbels and times of harvest of the seed crop, *Seed Science and Technology* 7, 169–178.

Gray, D. (1981) Are the plant densities currently used for carrot seed production too low? *Acta Horticulturae* 111, 159–165.

Gray, D. (1983) Improving the quality of vegetable seeds. *Span* 26(1), 4–6.

Gray, D. (1994) Large-scale seed priming techniques and their integration with crop protection. In Martin, T. (ed.) *Seed Treatment: Progress and Prospects.* British Crop Protection Council, Surrey, UK, pp. 353–362.

Gray, D. and Steckel, J.R.A. (1980) Studies on the sources of variation in plant weight in *Daucus carota* (carrot) and the implications for seed production techniques. In: Hebblethwaite, P.D. (ed.) *Seed Production.* Butterworths, London and Boston, pp. 474–484.

Gray, D. and Steckel, J.R.A. (1983) Some effects of umbel order and harvest date of the carrot seed crop on seed variability and seedling performance. *Journal of Horticultural Science* 58(1),73–83.

Gray, D. and Steckel, J.R.A. (1984) Variability of onion (*Allium cepa*) seed as influenced by temperature during seed growth. Annals of *Applied Biology* 104, 375–382.

Gray, D. and Steckel, J.R.A. (1985a) Parsnip (*Pastinaca sativa*) seed production: effects of seed crop plant density, seed position on mother plant, harvest date and method, and seed grading on embryo and seed size and seedling performance. *Annals of Applied Biology* 107, 559–570.

Gray, D. and Steckel, J.R.A. (1985b) Variation in flowering time as a factor influencing variability in seeding size in the subsequent carrot crop (Daucus carota L.). *Journal of Horticultural Science* 60, 77–81.

Gray, D. and Steckel, J.R.A. (1986a) The effects of seed-crop plant density, transplant size, harvest date and seed grading on leek (*Allium porrum* L.) seed quality. *Journal of Horticultural Science* 61 (3), 315–323.

Gray, D. and Steckel, J.R.A. (1986b) The effects of several cultural factors on leek (*Allium porrum* L.) seed production. *Journal of Horticultural Science* 61 (3), 307–313.

Gray, D. and Ward, J.A. (1985) Relationships between seed weight and endosperm char-acteristics in carrot. *Annals of Applied Biology* 106, 379–384.

Gray, D., Hulbert, S. and Senior, K.J. (1985c) The effects of seed position, harvest date and drying conditions on seed yield and subsequent performance of cabbage. *Journal of Horticultural Science* 60, 65–75.

Gray, D., Steckel, J.R.A. and Ward, J.A. (1985d) The effect of plant density, Harvest date and method on the yield of seed and components of yield of parsnip (*Pastinaca sativa*). *Annals of Applied Biology* 107, 547–558.

Gray, D., Steckel, J.R.A., Dearman, J. and Brocklehurst, P.A. (1988a) Some effects of temperature during seed development on carrot (*Daucus carota*) seed growth and quality. *Annals of Applied Biology* 112, 367–376.

Gray, D., Wurr, D.C.E., Ward, J.A. and Fellows, J.R. (1988b) Influence of post-flowering temperature on seed development and subsequent performance of crisp lettuce. *Annals of Applied Biology* 113, 391–402.

Gray, D., Steckel, J.R.A., Chrimes, J.R. and Davies, A.C.W. (1991) Density and harvest date effects on leek seed yields and quality. *Seed Science and Technology* 19(2), 331–340.

Greengrass, B. (1998) International union for the protection of new varieties of plants. In: Kelly, A.F. and George, R.A.T. (eds) *Encyclopaedia of Seed Production of World Crops.* Wiley, Chichester, UK, pp. 19–23.

Gregg, B. and Wannapee, P. (1985) Organization for effective management of a national seed programme for high- and low-input cropping systems. In: *Proceedings of the Seminar on Seed Production* Yaounde, Cameroon CTA and IAC Wageningen, The Netherlands, pp. 1–21.

Greven, M.M., McKenzie, B.A. and Hill, G.D. (1997) The influence of stress on yield, abortion and seed size of French beans (*Phaseolus vulgaris* L.). *Proceedings Annual Conference 27*, Agronomy Society of New Zealand, New Zealand, pp. 101–108.

Groot, S.P.C., Oosterveld, O., van der Wolf, J., Jalink, W.M., Langeraak, C.J. and van den Bulk, R.W. (2004) The role of ISTA and seed science in assuring organic farmers with high quality seeds. In: FAO (ed.) *First World Conference on Organic Seed Proceedings*. Published jointly by The International Federation for Organic Movements (IFOAM), The International Seed Federation (ISF) and FAO, pp. 9–12.

Grubben, G.J.H. (1976) *The Cultivation of the Amaranth as a Tropical Leaf Vegetable*. Royal Tropical Institute, Amsterdam, 207pp.

Grubben, G.J.H. (1977) *Tropical Vegetables and Their Genetic Resources*. IBPGR, Rome.

Haldemann C. (2008) ISTA and biotech/GM crops. In: *Seed Testing International*, ISTA News Bulletin No. 136 October 2008, 3–5.

Halmer, P. (1994) The development of quality seed treatments in commercial Practice. In: Martin, T.J. (ed.) *Seed Treatment: Progress and Prospects*. The British Crop Protection Council, Thornton Heath, pp. 363–374.

Hampton, J.G., TeKrony, D.M. and the Vigour Test Committee (1995) *Handbook of Vigour Test Methods*, 3rd edn. International Seed Testing Association, Zurich, 117pp.

Harland, S.C. (1948) Inheritance of immunity to mildew in Peruvian forms of *Pisum sativum. Heredity* 2, 263–269.

Harrington, J.F. (1960) The use of gibberellic acid to induce bolting and increase seed yield of tight-heading lettuce. *Proceedings of the American Society of Horticultural Science* 75, 476–479.

Harrington, J.F. (1963) Practical advice and instructions on seed storage. *Proceedings of the International Seed Testing Association* 28, 989–994.

Hawthorn, L.R. (1951) *Studies on Soil Moisture and Spacing for Seed Crops of Carrots and Onions*. USDA, Circular NO. 852.

Hawthorn, L.R. and Pollard, L.H. (1954) *Vegetable and Flower Seed Production*. The Blakiston Company Inc., New York, 626 pp.

Hawthorn, L.R., Bohart, G.E., Toole, E.H., Nye, W.P. and Levin, M.D. (1960) *Carrot Seed Production as Affected by Insect Pollination*. Bulletin 422, Agriculture Experiment Station, Utah State University.

Hedrick, U.P. (1972) *Sturtevant's Edible Plants of the World*. Dover Publications, New York, pp. 166–167.

Heiser, C.B. and Smith, P.G. (1953) The cultivated capsicum peppers. *Economic Botany* 7, 214–227.

Hepper, F.N. (1970) Bambara groundnut (*Voandzei subterranea*). *Field Crop Abstracts* 24, 1–6.

Heydecker, W. (1978) Primed seed for better crop establishment. *Span* 21(1),12–14.

Heydecker, W. and Gibbins, B.M. (1978) The 'priming' of seeds. *Acta Horticulturae* 83, 213–223.

Higgins, J. and Evans, J.L. (1985) Courgette variety performance trial technique and summary of results for 1982–1984 trials. *Journal of the National Institute of Agricultural Botany* 17, 145–150.

Holland, H. (1957) Classification and performance of red beet. *Annual Report of the National Vegetable Research Station* 16–42.

Holland, R.W.K. (1985) Techniques used in vegetable variety performance trials. Brassica crops. *Journal of the National Institute of Agricultural Botany* 17, 117–128.

Hollingsworth, D.F. (1981) The place of potatoes and other vegetables in the diet. In: Spedding, C.R.W. (ed.) *Vegetable Productivity*. Macmillan, London, pp. 6–13.

Hossain, M.M. (1996) Vegetable seed production through contract growers. In: Wee, M.M.B. (ed.) *Proceedings of a Workshop on Vegetable Crops Agribusiness*. Held at BARC, Dhaka, Bangladesh, May 1993. AVRDC, Taiwan, pp. 55–64.

Howard, H.W. (1976) Watercress. In: Simmonds, N.W. (ed.) *Evolution of Crop Plants*. Longman, London and New York.

Hrabovszky, J.P. (1982) Crop production in developing countries in 2000. In: Feistritzer, W.P. (ed.) *Seeds*. FAO, Rome, pp. 29–39.

IBPGR (1982) *Revised Wing Bean Descriptors*. International Board for Plant Genetic Resources, Rome.

IBPGR (1987) *Descriptors for Bambara Groundnut.* International Board for Plant Genetic Resources, Rome.

IBPGR (1988) *Descriptors for Eggplant.* International Board for Plant Genetic Resources, Rome.

IFOAM, ISF and FAO (2004a) *Proceedings of First World Conference on Organic Seed.* FAO, Rome, 193 pp.

IFOAM, ISF and FAO (2004b) *Report of The First World Conference on Organic Seed,* FAO, Rome,

IPGRI (1993) Diversity for Development. *The Strategy of the International Plant Genetic Resources Institute.* International Plant Genetic Resources Institute, Rome.

Innes, N.L. (1983) *Breeding Field Vegetables.* AVRDC, Taiwan, 34pp.

ISF (1974) *A 50 Year Old Family 1924–1974.* FIS, Nyon, 144 pp.

ISF (1994) *International Seed Trade Federation Rules and Usages for the Trade in Seeds for Sowing Purposes,* 12th edn. FIS, Nyon, 121 pp.

ISF (1996a) *International Seed Trade Federation Arbitration Procedure Rules for the Trade in Seeds for Sowing Purposes,* 5th edn. FIS, Nyon, 42 pp.

ISF (1996b) *FIS Strategic Plan.* FIS, Nyon, 48 pp.

ISTA (1987) *Handbook for Cleaning of Agricultural and Horticultural Seeds on Small-scale Machines* Part 1, International Seed Testing Association, Zurich, 126pp.

ISTA (1988) *Handbook for Cleaning of Agricultural and Horticultural Seeds on Small-scale Machines* Part 2, International Seed Testing Association, Zurich, 117pp.

ISTA (1995a) *Understanding Seed Vigour.* International Seed Testing Association, Zurich, 4pp.

ISTA (1995b) *Handbook of Vigour Test Methods,* 3rd edn. International Seed Testing Association, Zurich, 117pp.

ISTA (2007) *ISTA List of Stabilized Plant Names,* 5th edn. International Seed Testing Association, Zurich, 60pp.

ISTA (2009a) *International Rules for Seed Testing.* International Seed Testing Association, Zurich.

ISTA (2009b) *Handbook on Pure Seed Definitions,* 3rd edn. International Seed Testing Association, Zurich.

Izzeldin, H., Lippert, L.F. and Takatori, F.H. (1980) An influence of water stress at different growth stages on yield and quality of lettuce seed. *Journal of the American Society for Horticultural Science* 105, 68–71.

Jackson, J.C. (1985) The use of polythene tunnels for the production of vegetable seed in the UK. *Plasticulture* 66, 33–38.

Jandial, K.C., Samnotra, R.K., Sudan, S.K. and Gupta, A.K. (1997) Effect of steckling size and spacing on seed yield in radish (*Raphanus sativus* L.). *Environment and Ecology* 15 (1), 46–48.

Jeffs, K.A. (ed.) (1986) *Seed Treatment.* The British Crop Protection Council, 2nd edn. Thornton Heath, 332 pp.

Jenkins, J.A. (1948) The origin of the cultivated tomato. *Economic Botany* 2, 379–392.

Jianhua, Z., McDonald, M.B. and Sweeney, P.M. (1996) Soybean cultivar identification using RAPD. *Seed Science and Technology* 24, 589–592.

Johnson, A.G. (1956) Spring cabbage varieties. Annual *Report of the National Vegetable Research Station,* pp. 17–34

Johnson, A.G. (1958) How to increase top quality sprout yields. *Grower,* 1 November.

Johnson, A.G. and Haigh, J.C. (1966) The effect of intensity of selection during successive operations of seed multiplication on the field performance of Brussels sprouts. *Euphytica* 15, 365–373.

Johnson, D.T. (1968) The bambara groundnut, a review. *Rhodesia Agriculture Journal* 65, 1–4.

Jones, H.A. and Mann, L.K. (1963) *Onions and Their Allies.* Leonard Hill, London, 286 pp.

Jones, L.H. (1963) The effect of soil moisture gradients on the growth and development of broad beans (*Vicia faba* L.). *Horticultural Research* 3, 13–26.

Jones, R.A.C. (1982) Tests for transmission of four potato viruses through potato true seed. *Annals of Applied Biology* 100, 315–320.

Kachru, R.P. and Sheriff, J.T. (1992) Performance evaluation of axial-flow vegetable seed extractor. *Indian Journal of Agricultural Engineering* 2(1), 37–40.

Kelly, A.F. (1978) The role of seed programmes in agricultural development. In: Feistritzer, W.P. and Kelly, A.F. (eds) *Improved Seed Production*. FAO, Rome, pp. 13–24.

Kelly, A.F. (1988) *Seed Production of Agricultural Crops*. Longman, Harlow, pp. 156–157.

Kelly, A.F. (1994) *Seed Planning and Policy for Agricultural Production*. Wiley, Chichester, UK, 182 pp.

Kelly, A.F. and George, R.A.T. (1998a) Reserve seed stocks against major disasters. In: *Encyclopaedia of Seed Production of World Crops*, Wiley, Chichester, UK, pp. 147–149.

Kelly, A.F. and George, R.A.T. (1998b) International Agreement and National Legislation. In: *Encyclopaedia of Seed Production of World Crops*, Wiley, Chichester, UK, pp. 19–67.

Khan, A.A. (1992) Preplant physiological seed conditioning. In Janick, J. (ed.) *Horticultural Reviews*, Vol. 13. Wiley, New York, pp. 131–181.

Kinet, J.M. and Peet, M.M. (1997) Tomato In: Wien, H.C. (ed.) *The Physiology of Vegetable Crops*. CAB International, Wallingford and New York, pp. 207–258.

Kreyger, J. (1963) Equipment of importance for seed drying in Europe. *Proceedings of the International Seed Testing Association* 28(4) 793–826.

Krug, H. (1997) Environmental influences on development, growth and yield. In: Wien, H.C. (ed.) *The Physiology of Vegetable Crops*. CAB International, Wallingford, UK, pp. 101–180.

Lal, G. (1993a) Seed production techniques of tomato and brinjal in the tropics and subtropics. In: *Breeding of Solanaceous and Cole crops*. AVRDC, Taiwan. pp. 61–81.

Lal, G. (1993b) Seed production technology for tropical and subtropical cauliflower. In: *Breeding of Solanaceous and Cole crops*. AVRDC, Taiwan, pp. 268–279.

Lammerts van Bueren, E.T., van Soest, L.J.M., de Groot, E.C., Boukema, I.W. and Osman, A.M. (2005) Broadening the genetic base of onion to develop better-adapted varieties for organic farming systems. *Euphytica* 146, 125–132.

Laverack, G.K. and Turner, M.R. (1995) Roguing seed crops for genetic purity: a review. *Plant Varieties and Seeds* 8, 29–46.

Le Buanec, B. (1998) Fédédaration Internationale du Commerce des Semences (FIS)-International Seed Trade Association. In: Kelly, A.F. and George, R.A.T. (eds) *Encyclopaedia of Seed Production of World Crops*. Wiley, Chichester, UK, pp. 36–39.

Legro, R.J. (2004) Organic seed & coating technology: a challenge and opportunity. In: IFOAM, ISF and FAO (2004) *Proceedings of First World Conference on Organic Seed*, FAO, Rome, pp. 108–110.

Lesley, J.W. (1924) Cross pollination of tomatoes. *Journal of Heredity* 15, 233–235.

Lemonius, M. (1998) Asia and Pacific Seed Association (APSA). In: Kelly, A.F. and George, R.A.T. (eds) *Encyclopaedia of Seed Production of World Crops*. Wiley, Chichester, UK, pp. 39–40.

Lewis, D. (1979) *Sexual Incompatibility in Plants*. Studies in Biology No. 110. Edward Arnold, London. 60pp.

Liaw, H.L. (1982) *The Effect of Mineral Nutrition on Seed Yield and Quality in Phaseolus vulgaris*. MSc thesis, The University of Bath, Bath, UK.

Linnemann, A.R. and Azam-Ali, S. (1993) Bambara groundnut (Vigna subterranea). In: Williams, J.T. (ed.) *Underutilized Crops: Pulses and Vegetables*. Chapman and Hall, London, pp. 13–58.

Longden, P.C. (1975) Sugar beet seed pelleting. *ADAS Quarterly Review* 18,73–80.

Lookhart, G.L. and Wrigley C.W. (1995) Variety identification by electrophoretic analysis. In: Wrigley, C.W. (ed.) *Identification of Food Grain Varieties*. AACC, USA, pp. 55–72.

Louwaars, N.P. and van Marrewijk, G.A.M. (1996) *Seed Supply Systems in Developing Countries*. CTA, Wageningen, 135pp.

MacKay, D.C. and Eaves, C.A. (1962) The influence of irrigation treatments on yields and on fertilizer utilization by sweetcorn and snap beans. *Canadian Journal of Plant Science* 42 (2), 219–227.

MacLeod, J. (2007) Heritage vs commercial vegetable cultivars: are they equally important to gardeners? *The Garden*, October, p. 653.

Maguire, J.D., Kropf, J.P. and Steen, K.M. (1973) Pea seed viability in relation to bleaching. *Proceedings of the Association of Official Seed Analysts* 63, 51–58.

Maude, R.B. (1986) Treatment of vegetable seeds. In: Jeffs, K.A. (ed.) *Seed Treatment*, 2nd edn. The British Crop Protection Council, Thornton Heath, pp. 239–261.

Maude, R.B. and Keyworth, W.G. (1967). A new method for the control of seed-borne fungal disease. *Seed Trade Review* 19, 202–204.

Maude, R.B., Vizor, A.S. and Shuring, C.G. (1969) The control of fungal seed-borne diseases by means of a thiram seed soak. *Annals of Applied Biology* 64, 245–257.

McNaughton, I.H. (1976a) Swedes and rapes. In: Simmonds, N.W. (ed.) *Evolution of Crop Plants*. Longman, London and New York.

McNaughton, I.H. (1976b) Turnip and relatives. In: Simmonds, N.W. (ed.). *Evolution of Crop Plants*. Longman, London and New York.

Millar, A.A., Gardner, W.R. and Goltz, S.M. (1971) Internal water status and water transport in seed onion plants. *Agronomy Journal* 63, 779–784.

Morell, M.K., Peakall, R., Appels, R., Preston, L.R. and Lloyd, H.L. (1995). DNA profiling techniques for plant variety identification. *Australian Journal of Experimental Agriculture* 35, 801–819.

Murthy, N.S.R. and Murthy, B.S (1962) Natural cross pollination in chilli. *Andhra Agriculture Journal* 9(3), 161–165.

Musopole, E. (1995) Actionaid seed involvement. In: *Proceedings of Workshop on Improved On-farm Seed Production for SADC Countries*. FAO, Rome, pp. 47–49.

NIAB (2007) *Recommended Varieties for Organic Growing*. National Institute of Agricultural Botany, Cambridge, UK, 64pp.

Nieuwhof, M. (1969) *Cole Crops*. Leonard Hill, London, 353 pp.

Nitzany, F.E. (1960) Transmission of tobacco mosaic virus through tomato seed and virus inactivation by methods of seed extraction and seed treatments. *Ktavim Agricultural Research Station* 10, 63–67.

OECD (1995a) *List of Cultivars Eligible for Certification*. Organization for Economic Co-operation and Development, Paris, pp. 12–14.

OECD (1995b) *OECD Guidelines to Methods Used in Control Plot Tests and Field Inspection*. Organization for Economic Co-operation and Development, Paris.

OECD (1996) *OECD Scheme for the Varietal Certification of Vegetable Seed Moving in International Trade*. Organization for Economic Co-operation and Development, Paris.

Ogawa, T. (1961) Studies on seed production of onion 1, effects of rainfall and humidity on fruit setting. *Journal of Japan Society of Horticultural Science* 30, 222–232.

Olympio, N.S. (1980) *Influence of Certain Pesticides on the Yield and Quality of Lettuce, Rape and Tomato Seeds*. PhD Thesis, University of Bath, Bath, UK.

Oomen, H.A.P.C. and Grubben, G.J.H. (1978) *Tropical Leaf Vegetables in Human Nutrition*. Royal Tropical Institute, Amsterdam and Orphan Publishing Company, Willemstad, Curacao, 140 pp.

Opeña, R.T. and Chen, J.T. (1993) Production of F1 Hybrid Tomato Seeds. In: *Breeding of Solanaceous and Cole Crops*. AVRDC, Taiwan, pp. 56–67.

Opeña, R.T., Shanmugasundaram, S., Yoon, Jin-Young and Fernandez, G.C.J. (1987) *Crop Improvement Programme to Promote Vegetable Production in the Tropics*. Asian Vegetable Research and Development Center, Taiwan, 17 pp.

Opeña, R.T., Kuo, C.G. and Yoon, J.Y. (1988) *Breeding and Seed Production of Chinese Cabbage in the Tropics and Subtropics*, Technical Bulletin No. 17, Asian Vegetable Research and Development Center, Taiwan.

Panthee, D.R. and Subedi, P.P. (1998) *Effect of Threshing, Cleaning and Drying Period on Onion Seed Viability*. Working Paper, Lumle Agricultural Centre. Kaski, Nepal, 7 pp.

Parera, C.A. and Cantliffe, D.J. (1994) Presowing seed priming. *Horticultural Reviews* 16, 109–141.

Payne, R.C. (1993a) *Growth Chamber-Greenhouse Testing Procedures: Variety Identification*. International Seed Testing Association, Zurich.

Payne, R.C. (1993b) *Rapid Chemical Identification Techniques*. International Seed Testing Association, Zurich, 22 pp.

Perry, D.A. and Harrison, J.G. (1973) Causes and development of hollow heart in pea seed. *Annals of Applied Biology* 73(1), 95–101.

PGRO (1978) *Information Sheet Number 70*. Processors' and Growers' Research Organisation.

Pickett, A.A. (1998) Genetic quality. In: Kelly, A.F. and George, R.A.T. (eds) *Encyclopaedia of Seed Production of World Crops*. Wiley, Chichester, UK, pp. 71–79.

Pillbeam, C. and Hebblethwaite, P. (1994) The faba bean. *Biologist* 41 (4) 169–172.

Pooler, M.R. and Simon, P.W. (1994) True seed production in garlic. *Sexual Plant Production* 7(5), 282–286.

Poulos, J.M. (1993) Pepper breeding. In: *Breeding of Solanaceous and Cole crops*. AVRDC, Taiwan, pp. 85–121.

Pressman, E. (1997) Celery. In: Wien, H.C. (ed.) *The Physiology of Vegetable Crops*. CAB International, Wallingford, UK, pp. 387–407.

Priestley, D.A., Cullinan, V.I. and Wolfe, J. (1985) Differences in seed longevity at the species level. *Plant Cell and Environment* 8(8), 557–562.

Prince, S.D. (1980) Vernalization and seed production in lettuce. In: Hebblethwaite, P.D. (ed.) *Seed Production*. Butterworths, London, pp. 485–499.

Purseglove, J.W. (1972) *Tropical Crops, Monocotyledons*. Longman, London.

Purseglove, J.W. (1974) *Tropical Crops, Dicotyledons*. Longman, London.

Rabinowitch, H.D. (1990) Physiology of flowering. In: Rabinowitch, H.D. and Brewster, J.L. (eds) *Onions and Allied Crops*, Vol. 1. CRC Press, Boca Raton, Florida, pp. 113–134.

Reisch, W. (1952) *Variabilitätsstudien an Vicia faba* L.Z. Acker-u PflBau 94, 281–306.

Richardson, M.J. (1990) *An Annotated List of Seed-borne Diseases*, 4th edn. The International Seed Testing Association, Zurich.

Richardson, R.W. and Alvarez, E.L. (1957) Pollination relationships among vegetable crops in Mexico, natural cross pollination in cultivated tomatoes. *Proceedings of The American Society for Horticultural Science* 69, 336–371.

Rick, C.M. (1950) Pollination relations of *Lycopersicon esculentum* in native and foreign regions, *Evolution* 4, 110–122.

Riggs, T.J. (1987) Breeding F1 hybrid varieties of vegetables. In: Feistritzer, W.P. and Kelly, A.F. (eds) *Hybrid Production of Selected Cereal Oil and Vegetable Crops*. FAO, Rome, pp. 149–173.

Riggs, T.J. (1988) Breeding F1 hybrid varieties of vegetables. *Journal of Horticultural Science* 63 (3), 369–382.

Robani, H. (1994) Film-coating horticultural seed. *HortTechnology* 4, 104–105.

Roberts, E.H. (1972) Storage environment and the control of viability. In: Roberts, E.H. (ed.) *Viability of Seeds*. Chapman and Hall, London, pp. 14–58.

Roberts, E.H. and Roberts, D.L. (1972) Viability nomographs. In: Roberts, E.H. (ed.) *Viability of Seeds*. Chapman & Hall, London, pp. 417–123.

Roberts, E.H., Summerfield, R.J., Ellis, R.H., Craufurd, P.Q. and Wheeler, T.R. (1997) The induction of flowering. In: Wien, H.C. (ed.) *The Physiology of Vegetable Crops*. CAB International, Wallingford, UK, pp. 69–99.

Robinson, R.W. and Decker-Walters, D.S. (1997) *Cucurbits*. CAB International, Wallingford and New York, 226pp.

Rodenburg, C.M. (1958) The identification of lettuce varieties from the young plant. *Euphytica* 7, 241–246.

Rodenburg, C.M. and Huyskes, J.A. (1964) The identification of varieties of lettuce, spinach and Witloof chicory. *Proceedings of the International Seed Testing Association* 29 (4), 963–980.

Rosselló, J.M.E. and Fernández de Gorostiza, M. (1993) *Technical Guidelines for Field Variety Trials*. FAO Plant Production and Protection Paper 75, FAO, Rome.

Rowse, H.R. (1996a) Drum priming – A non-osmotic method of priming seeds. *Seed Science and Technology* 24, 281–294.

Rowse, H.R. (1996b) Drum priming – An environmentally-friendly way of improving seed performance. *Journal of the Royal Agricultural Society of England* 157, 77–83.

Sadik, S. (1982) A method for seed extraction. *True Potato Seed (TPS) Letter* 2 (3).

Salter, P.J. (1958) The effects of different water-regimes on the growth of plants under glass. IV Vegetative growth and fruit development in the tomato, *Journal of Horticultural Science* 33, 1–12.

Salter, P.J. (1978a) Techniques and prospects for 'fluid drilling' of vegetable crops. *Acta Horticulturae* 72, 101–108.

Salter, P.J. (1978b) Fluid drilling of pre-germinated seeds: progress and possibilities. *Acta Horticulturae* 83, 245–249.

Salter, P.J. and Goode, J.E. (1967) *Crop Responses to Water at Different Stages of Growth*. Commonwealth Agricultural Bureaux, Farnham Royal, England.

Sapirstein, H.D. (1995) Variety identification by digital analysis. In: Wrigley, C.W. (ed.) *Identification of Food Grain Varieties*, AACC, USA, pp. 91–130.

Schoen, J.F. (1993) *Laboratory Tests for Varietal Determination with Fungal Pathogens*. International Seed Testing Association, Zurich, 15 pp.

Schuler, R.E., Taitt, R. and Farris, G.E. (2005) Farming practices influence wild pollinator populations on squash and pumpkin. *Journal of Economic Entomology* 98, 790–795.

Scott, J.M. (1989) Seed coatings and treatments and their effects on plant establishment. *Advances in Agronomy* 42, 43–83.

Seth, J.N. and Dhaudar, D.G. (1970) Effects of fertilizers and spacings on the seed yield and quality in brinjal (Solanum melongena L.) variety Pusa Purple Long. *Progressive Horticulture* 1(4), 45–50.

Sgaravatti, E. and Beaney, J. (1996) *World List of Seed Sources*, 4th edn. FAO, Rome, 623 pp.

Shaw, C.W. (1889) *Market and Kitchen Gardening*. Crosby Lockwood, London, pp. 140–141.

Sikora, I. (1995) Opening ceremony statement. In: *Proceedings of Workshop on Improved On-farm Seed Production for SADC Countries*. FAO, Rome, pp. 21–22.

Silvey, V. (1978) Methods of analysing NIAB variety trial data over many sites and several seasons. *Journal of the National Institute of Agricultural Botany* 14, 385–400.

Smartt, J. (1976) *Tropical Pulses*. Longman, London, 348 pp.

Smartt, J. (1990) *Grain Legumes: Evolution and Genetic Resources*. Cambridge University Press, Cambridge.

Smith, B.M. and Jackson, J.C. (1976) The controlled pollination of seeding vegetable crops by means of blowflies. *Horticultural Research* 16, 53–55.

Smith, F.L. (1955) The effects of dates of harvest operations on yield and quality of pink beans. *Hilgardia* 24, 37–52.

Smith, J.S.C. and Smith, O.S. (1991) Restriction fragment length polymorphisms can differentiate among U.S. maize hybrids. *Crop Science* 31, 893–899.

Smith, J.S.C. and Smith, O.S. (1992) Fingerprinting crop varieties. *Advances in Agronomy* 47, 85–140.

Smith, P.G., Villalon, B. and Villa, P.L. (1987) Horticutural classification of peppers grown in the United States. *HortScience* 22, 11–13.

Sneep, J. (1958) The breeding of hybrid varieties and the production of hybrid seed of spinach. *Euphytica* 7, 119–122.

Soo, T.T. (1977) Local horticultural seed production. In: Chin, H.F. Enoch and Raja Harun, R.M. (eds) *Seed Technology in the Tropics*. University Pertaniun, Malaysia, pp. 169–174.

Sparenberg, H. (1963) The artificial drying of unthreshed seed crops. *Proceedings of the International Seed Testing Association* 28(4), 785–792.

Spencer, D.M. and Glasscock, H.H. (1953) Crook root of watercress. *Plant Pathology* 2, 19.

Sperling, L., Loevinsohn, M.E. and Ntabomvura, B. (1993) Rethinking the farmer's role in plant breeding: local bean experts and on-station selection in Rwanda. *Experimental Agriculture* 29, 509–519.

Steiner, J.J. and Akintobi, D.C. (1986) Effect of harvest maturity on viability of onion seed. *HortScience* 21, 1220–1221.

Still, D.W. and Bradford, K.G. (1998) Using hydrotime and ABA-time models to quantify seed quality of brassicas during development. *Journal of the American Society for Horticultural Science* 123(4), 692–699.

Summerfield, R., Roberts, E., Ellis, R., Qi, A., Craufurd, P. and Wheeler, T. (1997) Timing it right prediction of flowering. *Biologist* 44 (4), 412–416.

Takahashi, O. (1987) Utilization and seed production of hybrid vegetable varieties in Japan. In: Feistritzer, W.P. and Kelly, A.F. (eds) *Hybrid Production of Selected Cereal Oil and Vegetable Crops*. FAO, Rome, pp. 313–328.

Tan-Kim-Yong, U. and Nikopornpun, M. (1993) Seed production adoption and perception among Karen and Hmong farmers of Northern Thailand. In: Thomas, N. and Mateo, N. (eds) *Proceedings of a Workshop on Seed Production Mechanisms Held in Singapore*, November 1990. International Development Research Centre, Ottawa, Canada, pp. 48–61.

Taylor, A.G. (1997) Seed storage, germination and quality. In: Wien, H.C. (ed.) *The Physiology of Vegetable Crops*. CAB International, Wallingford, UK, pp. 1–36.

Taylor, A.G., Min, T.G. and Mallaber, C.A. (1991) Seed coating system to upgrade Brassicaceae seed quality by exploiting sinapine leakage. *Seed Science and Technology* 19(2), 423–434.

Taylor, A.G., Prusinski, J., Hill, H.J. and Dickson, M.D. (1992) Influence of seed hydration on seedling performance. *Hort-Technology* 2(3), 336–344.

Taylor, A.G., Churchill, D.B., Lee, S.S., Bilsland, D.M. and Cooper, T.M. (1993) Color sorting of coated Brassica seeds by fluorescent sinapine leakage to improve germination. *Journal of the American Society for Horticultural Sciences* 118(4), 551–556.

Thoday, P. (2007) *Two Blades of Grass, the Story of the Cultivation of Plants*. Thoday Associates, St Albans, Hertfordshire, UK.

Thomas, T.H., Gray, D. and Biddington, N.L. (1978) The influence of the position of the seed on the mother plant on seed and seedling emergence. *Acta Horticulturae* 83, 57–66.

Thompson, K.F. (1964) Triple-cross hybrid kale. *Euphytica* 13, 173–177.

Thompson, K.F. (1976) Cabbages, kales etc. In: Simmonds, N.W. (ed.) *Evolution of Crop Plants*. Longman, London and New York.

Tohme, J., Orlado Gonzalez, D., Beebe, S. and Duque, M.C. (1996) AFLP analysis of gene pools of a wild bean core collection. *Crop Science* 36, 1375–1384.

Tomlinson, J.A. (1957) Mosaic diseases of watercress. In: *Annual Report, National Vegetable Research Station*, 61.

Tonkin, J.H.B. (1979) Pelleting and other presowing treatments. *Research and Technology of Seeds* 4, 84–105.

Tripp, R. (2001) *Seed Provision and Agricultural Development*. James Curry Publishers, Oxford, 175 pp.

Tucker, W.G. and Gray, D. (1986). The effects of threshing and conditioning carrot seeds harvested at different times on subsequent seed performance. *Journal of Horticultural Science* 61(1) 57–70.

Umaerus, M. (1989) *True Potato Seed*. International Potato Center, Lima, Peru.

van den Berg, B.M. and Gabillard, D. (1994) Isoelectric focusing in immobilized pH gradient of melon (Cucumis melo L.) seed protein: methodical and genetic aspects, and application in breeding. *Electrophoresis* 15, 1541–1551.

van de Vooren, J.G. and van der Heijden, G.W.A.M. (1993) Measuring the size of French beans with image analysis. *Plant Varieties and Seeds* 6, 47–53.

Vernon, J. (1999) Vibrant colours of grain amaranths. *Journal of The Royal Horticultural Society* 124, 12–15.

Vos, P., Hogers, R., Bleeker, M., van de Lee, T., Hornes, M., Frijters, A., Pot, J., Peleman, J., Kuiper, M. and Zabeau, M. (1995) AFLP: a new technique for DNA fingerprinting. *Nucleic Acids Research* 23, 4407–4414.

Walkey, D.G.A., Cooper, V.C. and Crisp, P. (1974) The production of virus-free cauliflower by tissue culture. *Journal of Horticultural Science* 49, 273–275.

Wareing, P.F. and Phillips, I.D.J. (1981) *Growth and Differentiation in Plants*, 3rd edn. Pergamon, Oxford and New York, 343 pp.

Warren, D.E. (1997) Image analysis research at NIAB: chrysanthemum leaf shape. *Plant Varieties and Seeds* 10, 59–61.

Watts, L.E. (1955) Synonymy in lettuce varieties. *Annual Report of the National Vegetable Research Station*. pp. 16–36.

Watts, L.E. (1960) The use of a new technique in breeding for solidity in radish. *Journal of Horticultural Science* 35, 221–226.

Watts, L.E. (1980) *Flower and Vegetable Plant Breeding*. Grower Books, London, 182 pp.

Watts, L.E. and George, R.A.T. (1958) Classification and performance of varieties of radish. *Annual Report of the National Vegetable Research Station*, pp. 15–27.

Watts, L.E. and George, R.A.T. (1963) Vegetative propagation of autumn cauliflower. *Euphytica* 12, 341–345.

Waycott, W. (1993) Transition to flowering in lettuce: effect of photoperiod. *HortScience* 28, 530.

Weiring, D. (1964) The use of insects for pollinating brassica crops. *Euphytica* 13, 24–28.

Whyte, R.O. (1960) *Crop Production and Environment*. Faber and Faber, London.

Wien, H.C. (1997a) Peppers. In: Wien, H.C. (ed.) *The Physiology of Vegetable Crops*. CAB International, Wallingford, UK, pp. 259–293.

Wien, H.C. (1997b) Lettuce. In: Wien, H.C. (ed.) *The Physiology of Vegetable Crops*. CAB International, Wallingford, UK, pp. 479–509.

Wien, H.C. (1997c) Correlative growth in vegetables. In: Wien, H.C. (ed.) *The Physiology of Vegetable Crops*. CAB International, Wallingford, UK, pp. 181–206.

Wien, H.C. and Wurr, D.C.E. (1997) Cauliflower, broccoli, cabbage and brussels sprouts. In: Wien, H.C. (ed.) *The Physiology of Vegetable Crops*. CAB International, Wallingford, UK, pp. 511–552.

Weinberger, K. and Msuya, J. (2004) *Indigenous Vegetables in Tanzania*. AVRDC Technical Bulletin No. 31, 70 pp.

Williams, J.T. and Brenner D. (1995) Grain amaranth (*Amaranthus* spp.). In: Williams, J.T. (ed.) *Cereals and Pseudocereals*. Chapman & Hall, New York, pp. 129–186.

Wills, A.B. and North, C. (1978) Problems of hybrid seed production. *Acta Horticulturae* 83, 31–36.

Wilson, A. (1995) An NGO's experience in small scale seed production. In: *Proceedings of Workshop on Improved On-farm Seed Production for SADC Countries*. FAO, Rome, pp. 42–46.

Wilson, D.O. and McDonald, M.B. (1992) Mechanical damage in bean (*Phaseolus vulgaris* L.) seed in mechanized and nonmechanized threshing systems. *Seed Science and Technology* 20(3), 571–582.

Wilson, D.O. and Trawatha, S.E. (1991) Enhancement of bean emergence by seed moisturization. *Crop Science* 31, 1648–1651.

Wobil, J. (1994) Seed programme development in Swaziland. *Plant Varieties and Seeds* 7, 7–16.

Wolfe, D.W., Henderson, D.W., Hsiao, T.C. and Alvino, A. (1988) Interactive water and nitrogen effects on senescence of maize. I. Leaf area duration, nitrogen distribution and yield. *Agronomy Journal* 80, 859–864.

Wolfe, D.W., Azanza, F. and Juvik, J.A. (1997) Sweet corn. In: Wien, H.C. (ed.) *The Physiology of Vegetable Crops*. CAB International, Wallingford, UK, pp. 461–478.

Wurr, D.C.E., Jane, R., Fellows, D., Gray and Joyce R.A. Steckel (1986) The effects of seed production techniques on seed characteristics, seedling growth and performance of crisp lettuce. *Annals of Applied Biology* 108, 135–144.

Wurster, D.E. (1959) Air-suspension technique of coating drug particles. *Journal of the American Pharmaceutical Association* XLVIII, 451–454.

Yu, S., Cantliffe, D. and Nagata, R.T. (1998) Seed developmental temperature regulation of thermotolerance in lettuce. *Journal of the American Society for Horticultural Science* 123(4), 700–705.

General Index

Detailed information on a subject can be found under the names of specific vegetables as well as under the subject.

African eggplant 202
African spinach 270
agronomy 50–74
aid programmes 8, 10–12
alfalfa 38, 173
Alliaceae 42, 251–263
Amaranthaceae 270–274
Amaranthus 270–274
    cultivar description 271
    diseases 271, 274
    flowering 272
    harvesting 273
    isolation 272
    nutritional requirements 271
    pollination 272
    production areas 270
    production methods 271–272
    roguing 273
    soil requirements 271
    1000 grain weight 274
    yield 274
animal damage 75, 96, 101
anise 226
AOSA *see* Association of Official Seed Analysts
AOSCA *see* Association of Official Seed Certifying Agencies

aphids 54, 57, 136
Apiaceae 42, 55, 226–250
Asia and Pacific Seed Association (APSA) 16–17
Asian Vegetable Research and Development Center (AVRDC)
    crop improvement 18
    cultivar trials 26
    vegetable breeding material 9
    vegetable growing programmes 1
asparagus bean *see* winged bean
Association of Official Seed Analysts (AOSA) 108
Association of Official Seed Certifying Agencies (AOSCA) 31
Asteraceae 129–139
aubergine 52, 202, 214–218
    basic seed production 216
    crop rotation 215
    cultivar description 214–215
    diseases 215, 218
    drying 87–88
    extraction 81, 217
    F1 hybrids 216–217
    flowering 215
    harvesting 217
    nutritional requirements 216

aubergine (*continued*)
    pollination 215
    production methods 215–216
    roguing 216
    soil requirements 216
    1000 grain weight 218
    types 214
    yield 218
AVRDC *see* Asian Vegetable Research
        and Development Center

Bambara groundnut 181, 197–199
    diseases 199
    harvesting 198
    nutritional requirements 198
    production methods 198
    soil requirements 198
    1000 grain weight 199
    types 197–198
    yield 199
bamboo 264
basic seed, production methods 15, 32,
        55, 56, 62, 65, 114
    cleaning 85
    harvesting 78
    mosaic indexing 136
    storage 92
bean
    DNA profiling 30
    hydration 73
    longevity 93
    moisture content 100
    separating 84, 85
    types 181
            see also broad bean; French
                    bean; runner bean;
                    winged bean
bees *see* pollination by insects
beetroot 116–122
    basic seed production 118, 121
    cultivar description 117
    diseases 122
    drying 89
    flowering 119
    harvesting 121
    identification 29
    isolation 120
    nutritional requirements 117–118
    pollination 60, 116, 119–120
    production areas 116, 117

production methods 118–119
    roguing 62, 65, 118, 120–121
    root storage 118
    root-to-seed system 118–119
    seed forms 116–117
    seed-to-seed system 119
    selection 116
    soil requirements 117
    1000 grain weight 122
    threshing 121
    types 116
    vernalization 116
    yield 122
biennial crops 6, 50–51, 62, 64–65
    organic seed production 51, 67, 68
    water requirements 48
bird chilli *see* pepper
bird damage 54, 75
blowflies *see* pollination by insects
Brassicaceae see Cruciferae
brassicas 140–161
    agronomy 156, 159
    coated seeds 73
    cultivar trials 26
    hybrids 46
    hydration 73
    identification of cultivars 31
    separating 83
    shattering 75
    sorting 73
breeders' seed 32, 62, 64
    harvesting 78
    storage 92, 98
brinjal 214
broad bean 29, 181, 194–197
    cultivar description 194–195
    diseases 197
    flowering 195
    harvesting 196
    isolation 196
    nutritional requirements 195
    pollination 195–196
    production areas 194
    production methods 195
    roguing 196
    soil requirements 195
    1000 grain weight 196
    types 194
    water requirements 195
    yield 196
broccoli, sprouting *see* cauliflower

Brussels sprout 26, 140
    basic seed production 151
    cultivar description 141–142
    F1 hybrids 44
    pollination 42, 151
    production methods 143, 144
    roguing 62, 146
        see also cole crops
bush greens 270

cabbage 26, 140, 141
    basic seed production 149
    F1 hybrids 9
    harvesting 147–148
    production methods 143, 144
    roguing 146
    water requirements 47
        see also cole crops
cabbage, Chinese *see* Chinese cabbage
cages 42, 44, 56–57
cantaloupe melon *see* melon
cape gooseberry 202
caraway 226, 227
carrot 26, 55, 226, 227–237
    basic seed production 232, 237
    cultivar description 230–231
    debearding 235
    diseases 54, 237
    drying 89, 235
    F1 hybrids 44, 236–237
    flowering 227–228, 231–232
    harvesting 76, 78, 235
    identification 28
    isolation 232–233
    longevity 98
    nutritional requirements 233
    pollination 38, 232, 237
    production methods 227–230,
        236–237
    quality 227–228, 228–230
    roguing 234
    root-to-seed production 228,
        233–234, 237
    seed priming 72
    seed-to-seed production 65, 228,
        233, 234
    selection 62, 237
    separating 83, 85, 235
    shattering 235
    soil requirements 233

1000 grain weight 236
    threshing 235
    types 227
    vernalization 231
    yield 228–229, 235–236
casaba 172
catalogues 111
cauliflower 140
    basic seed production 144, 149–150
    cultivar description 141
    flowering 48
    nutritional requirements 142
    organic seed production 68
    production methods 143, 144
    roguing 146, 150
    selection 149–150
    vegetative propagation 150
        see also cole crops
CCD *see* colony collapse disorder
celeriac 226, 249–250
    cultivar description 249
    diseases 249
    production methods 249
    roguing 250
celery 26, 226, 227, 245–249
    agronomy 247–248
    basic seed production 247, 248
    blanching 245–246
    cultivar description 246
    diseases 246, 249
    flowering 247
    harvesting 248
    isolation 247
    nutritional requirements 246
    pollination 247
    production areas 247
    roguing 247, 248
    separating 248
    shattering 248
    soil requirements 246
    1000 grain weight 248
    types 245–246
    water requirements 246
    yield 248
Celosia 270–274
    agronomy 271–274
    cultivar description 271
    diseases 271, 274
    flowering 272
    harvesting 273
    isolation 272

Celosia (continued)
    nutritional requirements 271
    pollination 272
    roguing 273
    soil requirements 271
    1000 grain weight 274
    yield 274
cereals *see Gramineae*
certification schemes 16, 31–33, 54,
    55, 62, 110
certified seeds 32, 55, 59, 106, 136
chard *see* Swiss chard
Chenopodiaceae 116–128
cherry capsicum *see* pepper
chervil 226
chicory 129, 136–138
    agronomy 137
    diseases 137, 138
    identification 137
    1000 grain weight 137
    vernalization 136
    yield 137
chilli pepper *see* pepper
Chinese cabbage 44, 140, 141,
    151–152
    cultivar description 151
    diseases 149
    F1 hybrids 152
    nutritional requirements 151
    1000 grain weight 152
    water requirements 151
    yield 152
Chinese chives *see* onion
Chinese mustard 140, 151
Chinese pumpkin 162
Chinese spinach 270
chives *see* onion
classification of vegetables 3–4
cleaning 81–82
    see also separating
climatic factors
    FAO classification 50
    effect on pollination 41
    effect on storage potential 93, 97
cluster pepper *see* pepper
coated seeds 71, 73, 86
cock's comb *see Celosia*
cole crops 26, 140–151
    basic seed production 148–151
    diseases 141, 148–149
    F1 hybrids 17, 147

    flowering 141
    harvesting 147–148
    isolation 60, 146
    nutritional requirements 142
    pollination 38, 144–146
    production methods 143–145
    roguing 146
    selection 143
    shattering 147
    1000 grain weight 148
    threshing 147
    types 140–141
    water requirements 143
    yield 148
colony collapse disorder (CCD) 40–41
combine harvesting 79
Compositae see Asteraceae
coriander 226
cotton 38
courgette 26, 163
cover crops 54
cress *see* garden cress; watercress
crookneck squash 162, 178
crop protection chemicals 5, 65–66
    and organic seed production 51, 66
    and pollinating insects 38–40,
        41, 65
    seed treatments 9, 70, 71
crop rotation 4, 54–55, 69
cropping, protected 2, 9, 28, 42, 44,
    56–57
Cruciferae 6, 41, 42, 60, 140–161
cucumber 162, 168–172
    basic seed production 170
    cultivar description 168
    diseases 169, 172
    extraction 171
    flowering 169–170
    harvesting 171
    hybrids 44, 169, 170, 171
    isolation 170
    nutritional requirements 169
    pollination 170
    production areas 168
    production methods 169–170
    roguing 171
    separating 83
    soil requirements 169
    1000 grain weight 171–172
    types 168
    yield 171

Cucurbita pepo group 176–179
  cultivar descriptions 176
  diseases 179
  harvesting 178–179
  hybrids 177
  isolation 177
  nutritional requirements 176
  pollination 177
  production methods 176–177
  roguing 178
  soil requirements 176
  1000 grain weight 179
  yield 179
Cucurbitaceae 17, 80, 93,
  162–180
cucurbits 17, 80, 93, 162–180
  basic seed production 162
  drying 87, 88, 89
  flowering 162
  harvesting 77
  isolation 162
  longevity 93
  production areas 162, 179–180
  roguing 162
  types 162–163, 179–180
cultivars
  classification 27
  description, UPOV test
    guidelines 19, 281–282
  development 8–9, 17–18
  DUS testing 22, 27, 31, 32,
    64, 130
  growing-on tests 23, 35, 64
  identification methods
    chemical 29
    DNA profiling 30
    electrophoretic 29–30
    image analysis 30–31
    morphological 27–28
  lists of preferred 24–25
  maintenance 61–64
  naming 20–22, 27
  progeny testing 64
  registration 19, 22
  release 21–22
  trials
    design 25–27
    objectives 22–24
    organization 25
  and varieties 19–20
cumin 226, 227

dandelion 129
debearding 83
dehumidification of seed stores 99
demonstration of cultivars 2, 111
desiccants 78, 99, 101, 185, 229
dill 83, 226, 227
discard strip technique 61, 120, 121
disease
  alternative hosts 54, 56, 58,
    146, 202
  chemical control 65–66
  mosaic indexing 136
  organic seed production 55, 57, 67,
    68, 69, 70, 71
  and relative humidity 50
  resistance to 8–9, 28–29, 44
  seed-borne pathogens 33, 50, 55,
    56, 70, 106
  seed treatments 9, 70, 71
  separating diseased seeds 84, 85
  in storage 94, 95–96, 100, 101
  vectors 54, 57, 136
  and vegetative propagation 150
  see also pests
distribution 113–114
double-cross hybrids 46
drying 78, 87–89
  effect on storage potential 93, 94
DUS (distinctness, uniformity, stability)
  testing 22, 27, 31, 32, 64, 130
dwarf bean *see* French bean

earth bean *see* Bambara groundnut
eggplant *see* aubergine
egusi 162, 163
elephant garlic *see* leek
endive 129, 138–139
  agronomy 138–139
  cultivar description 138
  diseases 139
  isolation 139
  roguing 139
  1000 grain weight 139
  types 138
  yield 139
environment
  control for storage 94–101
  modification *see* microclimate
    modification
extraction *see* wet-extracted seeds

F1 hybrids 18, 22, 43–46
    breeding systems 44–45
    use by growers 45–46
faba bean 181
Fabaceae see Leguminosae
FAO *see* Food and Agriculture
        Organization
farming systems, vegetables in 4–5
fennel 226
fermentation 166, 179, 211
fertilization 37–43, 52
field bean 181
financial assistance for seed industry 8,
        10–11
fleshy fruit, seed extraction *see* wet-
        extracted seeds
Florence fennel 226
flowering 37
    synchronization 45
    water requirements 47–48
fluid drilling 72, 73
fluted pumpkin 162
fodder crops
    beans 194
    beet 60
    brassicas 46, 140, 141
    lucerne 38, 173
    peas 182
Food and Agriculture Organization
        (FAO) 8, 68
    climatic factors, classification 51
    Quality Declared Seed
            System 33–34
    Seed Improvement and
            Development
                Programme 10
    Seed Service 66
    soil conditions, classification 50
    World List of Seed Sources 14, 92
food security 5, 25, 92, 197
French bean 55, 181, 186–191
    basic seed production 189
    cultivar description 187
    diseases 191, 192
    flowering 48, 188
    harvesting 78, 190–191
    identification 28
    isolation 189
    nutritional requirements 187
    pollination 38, 188–189
    production areas 186

production methods 187–191
    roguing 189
    selection 189
    shattering 190
    1000 grain weight 191
    threshing 191
    types 186–187
    water requirements 187
    yield 191
fungi *see* disease

garden cress 140
garlic 251
gene banks 92–93
generations of seeds 32
genetic engineering *see* genetically
        modified organisms
genetically modified organisms
        (GMOs) 46–47, 60, 61,
        72, 269
    International Seed Testing
            Association role 107–108
germination
    and age at harvesting 77
    desiccants 78
    and moisture content 95, 100
    primed seeds 72–73
    synchronization 72–73
    testing 106–107
germplasm collections *see* gene banks
GMOs *see* genetically modified
        organisms
Goa bean *see* winged bean
gombo *see* okra
GOT *see* growing-on tests
gourd 170, 179–180
government roles
    in certification 32
    in cultivar choice 24
    in cultivar trials 25
    in quality control 34
    in seed industry 6–7, 8, 14
grading 9, 84, 86
grain legumes 3, 181
Gramineae 264–269
great-headed garlic *see* leek
green bean *see* French bean
green soko *see* celosia
greenhouses 28, 42, 44, 57
ground bean *see* Bambara groundnut

groundnut 98, 181, 197–199
growing-on tests (GOT) 23, 35, 64
growth chambers 28

Hamburg parsley *see* parsley
haricot bean *see* French bean
Harrington's rules for storage
    potential 95
harvesting
    desiccants 78
    dry seed 77, 78
    effect on storage potential 94
    fleshy fruits 77
    methods 77–79
    stages 76–77
    wet-extracted seeds 77
Hayek watercress 140, 159–160
health and vegetables 1, 2, 4, 24, 251
herbicides 5, 41, 65–66
hot water treatment 71
hybrid cultivars 5, 43, 46
    see also F1 hybrids
hydration treatments 72–73
hygiene in storage 101

identification of cultivars 27–31
IFOAM *see* International Federation of
    Organic Movements
ISF *see* International Seed Federation
Indian dill 226
Indian lettuce 129
Indian mustard 140
Indian spinach 125
indigenous plants 4, 5
insecticides and pollinating
    insects 38–40, 65
insects
    pollination by 38–43, 44, 52, 56,
        57, 65
    see also pests
intercropping 245
International Code of Botanical
    Nomenclature 20, 21
International Federation of Organic
    Movements (IFOAM) 66, 68
International Plant Genetic Resources
    Institute (IPGRI) 18
International Seed Federation
    (ISF) 15–16, 66, 68

International Seed Testing Association
    (ISTA) 16
    and GMOs 47, 107–108
    publications 20, 28–29, 83, 89,
        106, 107
    and seed testing 107–108
International Union for the Protection
    of New Varieties of Plants
    (UPOV) 19
    Test Guidelines 19, 281–282
IPGRI *see* International Plant Genetic
    Resources Institute
irrigation 47–48, 51, 93
isolation 56, 58–61
    spatial 52, 59–60
    temporal 59, 91
ISTA *see* International Seed Testing
    Association

Japanese bunching onion *see* onion

kale 140
    see also cole crops
kidney bean *see* French bean
kohlrabi 140
    cultivar description 142
    production methods 143–144
    roguing 144, 146
        see also cole crops
kurrat *see* leek

labelling 105, 110
    stored seeds 102
    truth in 34–35
lady's finger *see* okra
lagos spinach *see* celosia
leaf beet *see* Swiss chard
leek 26, 251, 259–263
    agronomy 260
    basic seed production 260, 261,
        262
    cultivar description 259–260
    diseases 258, 263
    drying 89, 262–263
    extraction 263
    F1 hybrids 259
    harvesting 78, 262
    isolation 262

leek (*continued*)
    longevity 93
    nutritional requirements 260
    pollination 262
    production areas 259, 261
    roguing
        root-to-seed 261
        seed-to-seed 261
    root-to-seed production 261
    seed priming 72
    seed-to-seed production 260–261
    soil requirements 260
    1000 grain weight 263
    types 259
    yield 263
legal requirements 14
    and costs 113
    cultivar release 21, 22
    genetically modified organisms 47
    international seed trading 15–16, 17
    labelling 110
    organic seed production 66–67, 72
    plant breeding 19
    of QDS system 34
    and quality 15, 106, 108
    record keeping 104
    zoning 60
legumes 181–201
    drying 89
    moisture content 100
    multiplication rates 32
    in organic farming 55
    production areas 51
    types 3
    water requirements 47
Leguminosae 3, 32, 55, 99, 181–201
lettuce 26, 129–136
    agronomy 131–132
    basic seed production 136
    classification 130
    cultivar description 131
    development 130, 134
    diseases 54, 57, 131, 133,
        136, 137
    drying 89
    flowering 132, 134
    harvesting 78, 134
    identification 28, 130
    isolation 132–133
    longevity 98
    mosaic indexing 136

    nutritional requirements 131
    pollination 38, 132
    pre-cleaning 135
    roguing 130, 131, 133, 134, 136
    seed stalk emergence 133–134
    selection 130, 131, 133, 134
    separating 84, 85
    shattering 75
    soil requirements 131
    1000 grain weight 135
    types 129–130
    water requirements 47, 131–132
    yield 135
lodging 75–76
longevity
    natural 93
    and storage conditions 95, 98, 101
    in vapour-proof containers 101
loofah 180
lucerne 38, 173

Madagascar groundnut *see* Bambara
    groundnut
maize 264, 266
    identification 30
    moisture content 100
    shelling 83
    types 264
    zoning 61
Malabar spinach 125
Malvaceae 274–278
mange tout 182
mangel-wurzel *see* mangold
mangold 60, 116, 120, 122
marrow 93, 163, 170
    see also Cucurbita pepo group
melon 162, 172–175
    cultivar description 173
    diseases 175
    harvesting 174–175
    isolation 174
    nutritional requirements 173
    pollination 174
    production areas 172
    production methods 173–174
    roguing 174
    soil requirements 173
    1000 grain weight 175
    types 172–173
    yield 175

melon pear 202
microclimate
    after lodging 76
    modification 51–54
    for pollination 38
moisture content 76, 87
    measuring 89, 107
    stored seeds 93, 95–96, 97, 98,
        99–101
monitoring seed quality 105–108
multiplication 22, 105
    and isolation 59
    rate 14, 32, 91, 92, 100
    and selection 62, 63, 64, 65
muskmelon 81, 172
mustard
    black 140
    Chinese 140, 151
    Indian 140
    white 140

naming cultivars 20–22, 27
naranjillo 202
National Designated Authorities
        (NDA) 32
nutrition
    in flowering 48
    of mother plants 49, 65
    organic seed production 69
    and storage potential 93

OECD *see* Organisation for Economic
        Co-operation and Development
oilseed crops 97, 140, 181
okra 274–278
    agronomy 275
    cultivar description 275
    diseases 278
    extraction 277
    flowering 276
    harvesting 277
    isolation 276
    nutritional requirements 275–276
    pollination 276
    production areas 274–275
    roguing 276
    soil requirements 275
    1000 grain weight 278
    yield 277, 278

'on farm' seeds *see* saved seeds
onion 26, 251, 252–259
    basic seed production 253, 254, 256
    bulb-to-seed production 253–254
    cleaning 258
    cultivar description 252–253
    diseases 68, 253, 258
    drying 89
    F1 hybrids 44, 57, 256
    flowering 254–255
    harvesting 78, 257
    identification 28, 31
    isolation 255
    longevity 93, 96, 98
    nutritional requirements 253
    organic seed production 68
    pollination 38, 42, 255
    production areas 252
    production methods 253–254
    use of protective environments 57
    roguing 254
        bulb-to-seed 255–256
        seed-to-seed 255
    seed priming 72
    seed-to-seed production 65, 253,
        254
    selection 62, 254
    separating 258
    shattering 75, 257
    soil requirements 253
    1000 grain weight 259
    threshing 257
    types 252
    water requirements 253
    yield 259
organic seed production 66–69
    biennial crops 51, 68
    crop rotation 55, 69
    disease control 55, 56, 57, 67, 68,
        69, 70, 71–72
    and F1 hybrids 46
    legal requirements 66–67, 72
    marketing considerations 67–68
    principles 66, 69
    record keeping 104
Organisation for Economic Co-operation
        and Development (OECD) 16, 31
organizations
    supporting seed industry 8, 10–12,
        16–17
    supporting seed trading 15–17

packing 93, 108–110
   vapour-proof containers
      99–101, 108
packing character 89
pak-choi 140, 151
paprika *see* pepper
parasitic weeds 56
parsley 242–245
   agronomy 243–244
   cultivar description 243
   diseases 245
   harvesting 244
   intercropping 245
   isolation 243
   pollination 243
   production areas 242
   roguing 244
   root-to-seed production
      243–244
   seed-to-seed production 243
   1000 grain weight 244
   types 242–243
   yield 244
parsnip 55, 226, 238–242
   agronomy 238–239
   basic seed production 239, 240
   cultivar description 238
   diseases 241, 242
   F1 hybrids 44
   harvesting 241
   nutritional requirements 239
   pollination 240
   production areas 238
   roguing 240–241
   root-to-seed production 239,
      240, 241
   seed-to-seed production 239,
      240, 241
   selection 239, 240, 241
   separating 241
   shattering 241
   soil requirements 239
   1000 grain weight 242
   threshing 241
   types 238
   vernalization 238
   yield 241, 242
Participatory Plant Breeding
   (PPB) 111–112
pathogens *see* disease
pe-tsai 140, 151
pea 15, 181, 182–186

basic seed production 184
cultivar description 182
diseases 50, 182, 186
flowering 184
harvesting 76, 78, 185
identification 29
moisture content 76, 100
nutritional requirements 65,
   182–183
pollination 38, 42, 184
production areas 182
production methods 182–184
roguing 184
separating seeds 84, 85
soil requirements 182–183
1000 grain weight 185
types 182
water requirements 183–184
yield 185
peanut *see* groundnut
pelleted seeds 9, 70–71, 86
   hydration 73
pepino 202
pepper 202, 218–222
   basic seed production 220–221
   cleaning 221
   cultivar description 219
   diseases 219, 222, 223
   drying 87–88, 221
   extraction 221
   F1 hybrids 221
   harvesting 77–78
   nutritional requirements 219–220
   pollination 219, 220
   production areas 218–219
   production methods 220–221
   roguing 220–221
   seed-borne diseases 50
   shelter belts 52
   soil requirements 219
   1000 grain weight 222
   types 218–219
   water requirements 220
   yield 221–222
pesticides *see* crop protection chemicals
pests
   alternative hosts 54, 56, 58
   natural predators 57
   seed damage 85, 106
   seed treatments 70
   in storage 94, 95–96, 101
   vectors of disease 54, 57, 136

pimento pepper 77
plant breeding 17–19
     hybrid cultivars 43–46
     rights of breeders 18–19, 22, 45
     screening trials 22–23
plant density 55–56
plant nomenclature *see* taxonomy
Poaceae see Gramineae
pollen beetles 41, 146
pollination 37–43
     climatic factors 41
     and crop rotation 54
     and field shape 61
     by insects 38–43, 44, 52, 65
          bees 38–41, 42, 44, 52
          blowflies 42, 44
          in enclosed structures 42,
               56, 57
     and isolation 58–61
     microclimate for 38, 76
     open 22, 44, 45, 63
     self-pollination 38, 42, 44, 62
     of synthetic hybrids 46
     by wind 37–38, 52, 56
          discard strip technique 61,
               120, 121
polythene tunnels 42, 44, 57
potato 202, 222–224
     agronomy 224
     cultivar description 223
     diseases 223, 224
     extraction 224
     1000 grain weight 224
     true potato seed 223
     yield 224
PPB *see* Participatory Plant Breeding
pre-basic seed 32, 55, 56–57, 62
pre-harvest factors 75–77, 93
pre-sowing treatments 70–73
precision drilling 5, 9, 70, 86, 110,
     117, 118
price components 113
processing seeds 81–90
     cleaning 81–82
     effect on storage potential 94
     management 86
     objectives 81
     rules 86
     separating 83–86, 148, 166,
          179, 211
     upgrading 56, 73, 83–86
     winnowing 82

production areas 14–15, 50
production methods *see* agronomy
progeny testing 64
pulses *see* legumes
pumpkin 81, 162, 163, 170
     see also Cucurbita pepo group
purity of seeds 28, 31, 32, 33, 54–55,
     58–61, 61–64, 72
     genetic 106, 114
     mechanical 106, 114
     tolerance limits 58–59

QDS *see* Quality Declared Seed System
quail grass *see Celosia*
quality
     attributes 106–107, 114
     deterioration during
          distribution 113–114
     importance of 8, 15
     postharvest assessment
          techniques 64
     pre-harvest factors 75–77
     in storage 94
          see also quality control
quality assurance 14, 31, 33, 107, 110
quality control 14, 15, 31, 105–108
     and QDS system 34
     and truth in labelling 34–35
          see also certification schemes
Quality Declared Seed System
     (QDS) 33–34
quinoa 116

radish 140, 152–155
     basic seed production 154
     cultivar description 152–153
     diseases 155
     flowering 154
     harvesting 78, 155
     isolation 154
     Japanese white 153
     nutritional requirements 153
     production methods 153
     roguing 152, 154
     1000 grain weight 155
     yield 155
rakkyo 251
record keeping 104–105
red beet *see* beetroot
refrigeration in storage 98

relative humidity
  and disease 50
  and storage 93, 94–96, 98–100
resistance to disease 8–9, 28–29, 44
ripening 75, 76, 87
roguing 62
  biennial crops 65
  efficiency guidelines 63
root celery *see* celeriac
root parsley *see* parsley
root-to-seed production 65
rotation of crops 4, 54–55, 69
runner bean 28, 181, 191–194
  agronomy 192
  basic seed production 192
  cultivar description 192
  diseases 192, 194
  flowering 193
  harvesting 194
  isolation 59, 193
  pollination 193
  roguing 193
  1000 grain weight 194
  types 191
  water requirements 193
  yield 194
'running out' 62
rutabaga 155–156

salsify 129
saved seeds 1, 6, 17, 45, 113
scalping 82
screening *see* trials of cultivars
seed-borne pathogens *see* disease
seed companies *see* seed industry
seed form modification 70–71
seed health, assessment 106
Seed Improvement and Development
    Programme (SIDP) 10
seed industry 1
  in developing countries 6–7, 8
  development 6–8, 10–14, 51
  financial assistance for 8, 10–11
  government involvement 6–7, 8, 14
  international production 14–17
  marketing 114
  organizations supporting 8, 10–12
  price levels 113
  promotion of products 110–111
  public relations 112–113

role in national economies 8–9
technical assistance for 8, 11–12,
    17–19
seed priming 72–73
seed programmes, assessing
    requirements 12–14
seed stores 96–103
seed-to-seed production 64–65
selection 62, 65, 112
  by growers 1, 6, 17, 22
  negative 63
  positive 56, 62
  pressures 22, 63
self-sufficiency 2, 5
separating 83–86, 148
  fermentation method 166,
    179, 211
  see also cleaning
shallot *see* onion
shattering 51, 75, 77
shelter crops 52
SIDP *see* Seed Improvement and
    Development Programme
soil conditions 51
  erosion 52
  FAO classification 50
  for mother plants 65
Solanaceae 48, 80, 202–225
solanaceous crops 202–225
  drying 87–88
  water requirements 48
sowing distances 55–56
soybean 30, 181
spacing of plants 55–56
spinach 44, 116, 125–128
  basic seed production 126
  cultivar description 125
  diseases 127, 128
  flowering 126
  harvesting 127
  hybrids 127
  isolation 126
  nutritional requirements 126
  pollination 38, 126
  production areas 51, 125
  production methods 126
  roguing 126–127
  soil requirements 126
  1000 grain weight 127
  threshing 127
  yield 127

spinach beet *see* Swiss chard
spring cabbage 149
sprouting broccoli *see* cauliflower
squash 88, 162, 170
    see also Cucurbita pepo group
standard seeds 32, 35
stecklings 65
storage 91–103
    environment 95–99
    period 94
    potential, factors affecting 93–99
        see also packing
    pre-harvest considerations 93–94
    reasons for 91–93
    security 105
    seed stores 96–99, 101–102
    vapour-proof containers
        99–101, 108
subsistence farming 2, 23, 24, 198
    crops grown 197, 199, 270
    and hybrid cultivars 5, 46
sugarbeet 52, 60, 116, 117,
    120, 122
sugarcane 264
sunflower 38, 53
swede 140, 156
    see also turnip
sweet corn 26, 38, 44, 264–269
    cultivar description 265
    diseases 268–269
    drying 89
    flowering 266
    harvesting 78, 267
    hybrids 46, 266–267
    isolation 266
    longevity 93
    moisture content 100
    nutritional requirements 265
    pollination 266
    production methods 265–266
    roguing 266
    separating 83, 267
    shelling 83, 267, 268
    as shelter crop 53
    soil requirements 265
    1000 grain weight 267
    water requirements 265–266
    yield 267
    zoning 61
sweet melon 162
sweet pepper *see* pepper

Swiss chard 116, 120, 122–125
    agronomy 123
    basic seed production 123
    cultivar description 123
    diseases 122, 125
    harvesting 124–125
    isolation 123
    nutritional requirements 123
    pollination 60
    production areas 122–123
    roguing 124
    soil requirements 123
    1000 grain weight 125
    threshing 125
    yield 125
synthetic hybrids 46

taxonomy 3, 4, 19–20, 29
technical assistance for seed industry 8,
    11–12, 17
temperature
    and storage 93, 95, 98
tests
    for certification 33
    for seed quality 106–108
        see also DUS testing;
            growing-on tests;
            progeny testing; trials
            of cultivars
threshing 78, 79–80, 94
tobacco 202
tomato 202, 203–214
    agronomy 204–205, 208–209
    cleaning 212–213
    crop rotation 207
    cultivar description 203–204
    diseases 204, 207, 211, 212,
        213–214
    drying 87–88, 89, 213
    extraction 81, 209–213
    F1 hybrids 203, 208–209
    flowering 206
    harvesting 55, 77, 203, 209
    identification 30
    isolation 207
    nutritional requirements
        205–206
    pollination 38, 42, 206–207
    roguing 207–208
    seed-borne diseases 50

tomato (*continued*)
    separating 83, 210–212
    soil requirements 205
    1000 grain weight 213
    types 203
    water requirements 48
    yield 213
topping 118, 157
TPS *see* true potato seed
trading, international 14–17
training 11–12, 17
tree tomato 202
trials of cultivars 22–27
    see also DUS testing; growing-on
        tests; progeny testing
triple-cross hybrids 46
true potato seed (TPS) 223
trueness to type 54, 61–64,
    106, 114
truth in labelling 34–35
turnip 60, 140
    basic seed production 157, 158
    cultivar description 156
    diseases 149, 156
    flowering 157–158
    harvesting 159
    isolation 157–158
    nutritional requirements 157
    pollination 158
    production methods 157
    roguing 158
    1000 grain weight 159
    yield 159
turnip rooted celery *see* celeriac

Umbelliferae 226–250
United Nations 8, 10, 18
upgrading 56, 73, 83–86
UPOV *see* International Union for the
    Protection of New Varieties of
    Plants

vapour-proof containers 99–101, 108
variety trials *see* trials of cultivars
varroa mite 40
vegetables
    classification 3–4
    in farming systems 4–5

health role 1, 2, 4, 24, 251
    new introductions 9
    preservation 5, 9
    production for sale 2, 5
    range of 3
    social development role 1, 2, 5
vegetative propagation 8, 150, 222,
    251, 279–280
vernalization 6, 37, 51
viability 106–107
    and storage 94, 95
vigour 107
    and storage 94
vine crops *see* Cucurbitaceae

watercress 140, 159–160
    agronomy 159
    flowering 159–160
    harvesting 160
    isolation 160
    pollination 159–160
    roguing 160
    1000 grain weight 160
watermelon 53, 162, 163–167, 170
    agronomy 164
    basic seed production 164
    cultivar description 163
    diseases 163, 167
    drying 166–167
    extraction 81, 165–166
    flowering 164
    harvesting 165–166
    isolation 165
    nutritional requirements 164
    pollination 164–165
    production areas 163
    roguing 165
    soil requirements 164
    1000 grain weight 167
    yield 167
weed control 5, 56, 65–66, 69,
    132, 143
Welsh onion *see* onion
wet-extracted seeds
    drying 87–88, 213, 221
    extraction 80–81, 83, 165–166,
        178–179, 209–213, 217,
        221
    harvesting 77, 171

white soko *see* celosia
windbreaks 52–54
winged bean 181, 199–200
    agronomy 200
    cultivar description 199
    diseases 200
    harvesting 200
    isolation 200
    roguing 200
    1000 grain weight 200
    yield 200
winnowing 82

winter melon 172
winter squash 162, 163, 178
witloof chicory *see* chicory

yield
    harvest times 76–77
    pre-harvest problems 75–77
    and water availability 47–48

zoning 60–61, 120, 146

Index of Species

Abelmoschus esculentus 274
Acremonium strictum 268
Albugo candida 149
Alliaceae 42, 251–263
Allium spp. 60
Allium ampeloprasum 259
Allium ampeloprasum var.
     porrum 251, 259
Allium ascalonicum 251
Allium cepa 20, 38, 93, 251, 252–259
Allium chinense 251
Allium fistulosum 251
Allium porrum 93, 251, 259–263
Allium sativum 251
Allium schoenoprasum 251
Allium tuberosum 251
Alternaria spp. 223
Alternaria alternata 122, 192, 218
Alternaria alternata f. sp.
     lycopersici 214
Alternaria amaranthi 274
Alternaria brassicae 149
Alternaria brassicicola 149, 155
Alternaria cichorii 138, 139
Alternaria cucumerina 172
Alternaria dauci 137, 237, 242,
     245, 249
Alternaria porri 258
Alternaria radicina 237, 242, 245, 249
Alternaria solani 214
Amaranthaceae 270–274

Amaranthus caudatus 270
Amaranthus cruentus 270
Amaranthus dubius 270
Amaranthus hypochondriacus 270
Amaranthus spinosa 272
Amaranthus tricolor 270
Anethum graveolens 226
Anethum sowa 226
Anthriscus sylvestris 226
Apiaceae 42, 55, 226–250
Apium dulce 226
Apium graveolens var. *dulce* 226,
     245–249
Apium graveolens var. *rapaceum* 226,
     249–250
Arabis mosaic virus 122, 137
Arachis hypogea 181
Ascochyta spp. 186, 192
Ascochyta abelmoschi 278
Ascochyta boltshauseri 192
Ascochyta fabae 197
Ascochyta phaseolorum 192
Ascochyta pisi 182, 186
Aspergillus spp. 192
Asteraceae 129–139
Atropa belladonna 202

Basella alba 125
Bean common mosaic virus 187
Bean yellow mosaic virus 182

Benincasa hispida 179
Beta spp. 60, 78
Beta vulgaris 29, 120, 122, 123
Beta vulgaris ssp. *cycla* 116, 122
Beta vulgaris ssp. *esculenta* 20, 116
Beta vulgaris ssp. *maritima* 116
Beta vulgaris var. *vulgaris* 122
Botrytis spp. 278
Botrytis allii 68, 253, 258
Botrytis byssoidea 258
Botrytis cinerea 133, 138, 249, 258
Botrytis fabae 197
Braconidae 41
Brassica spp. 31, 46, 73, 75, 140
Brassica campestris 60, 141
Brassica campestris ssp. *chinensis*
        140, 151
Brassica campestris ssp.
        pekinensis 140, 151
Brassica campestris ssp. *rapifera* 140,
        156–159
Brassica napus var. *napobrassica*
        140, 156
Brassica nigra 140
Brassica oleraceu 17, 38, 60, 140–141
Brassica oleracea var. *acepha* 140
Brassica oleracea var. *botrys* 140, 141
Brassica oleracea var. *capitata* 20, 140
Brassica oleracea var. *gemmifera* 140,
        141–142
Brassica oleracea var. *gongyloides* 140
Brassica oleracea var. *italica* 140
Brassicaceae see Cruciferae
Bremia lactucae 131

Capsicum spp. 218–222
Capsicum annuum 202, 218
Capsicum frutescens 77, 202, 218
Carrot red leaf virus 237
Carum carvi 226
Cauliflower mosaic virus 150
Cavariella aegopodii 54
Celosia argentea 270–274, 271
Celosia globerosa 270
Celosia insertii 270
Celosia leptostachya 270
Celosia pseudovirgata 270
Celosia trigyna 270
Centrospora acerina 241
Cercospora beticola 122

Cercospora canescens 192
Cercospora capsici 223
Cercospora carotae 237
Cercosporidium punctum 245
Chenopodiaceae 116–128
Chenopodium album 146
Chenopodium quinoa 116
Cherry leaf roll virus 192
Chicory yellow mottle virus 138
Choanephora cucurbitarum 271, 278
Cichorium endiva 129, 138–139
Cichorium intybus 129, 136–138
Citrullus lanatus 162, 163–167, 174
Citrullus vulgaris 162
Cladosporium allii 258
Cladosporium cladosporiodes f. sp.
        pisicola 186
Cladosporium cucmerinum 168,
        175, 179
Cladosporium fulvum 204
Cladosporium variabile 128
Clavibacter michiganensis 214
Cochliobolus carbonum 268
Cochliobolus heterostrophus 268
Cochliobolus spicifer var.
        melongenae 218
Coleoptera 232
Colletotrichum spp. 163, 167
Colletotrichum capsici 223
Colletotrichum circinans 258
Colletotrichum dematium f.
        spinaciae 122, 128
Colletotrichum graminicola 268
Colletotrichum lagenarium 172
Colletotrichum lindemuthianum
        192, 197
Colletotrichum melongena 218
Colletotrichum piperatum 223
Colletotrichum pisi 186
Colletotrichum spinaciicola 128
Compositeae see Asteraceae
Convolvulaceae 222
Coriandrum sativum 226
Corynebacterium michiganensis 211
Corynespora melonis 168
Cruciferae 6, 41, 42, 60, 140–161
Cucumber green mottle mosaic
        virus 172
Cucumber mosaic virus 125, 168, 172,
        175, 179, 192, 200, 214, 223
Cucumis anguria 179

Cucumis melo 162, 170, 172–175
Cucumis sativus 162, 168–172, 174
Cucurbita maxima 162
Cucurbita mixta 163
Cucurbita moschata 163
Cucurbita pepo 26, 93, 163
Cucurbitaceae 17, 80, 93, 162–180
Cuminum cyminum 226
Curtobacterium flaccumfaciens pv.
    betae 122
Cuscuta spp. 56
Cyphomandra betacea 202

Datura stramonium 202
Daucus carota 38
Daucus carota ssp. *sativus* 226
Diaporthe phaseolorum 223
Didymella spp. 163, 167
Didymella bryoniae 179
Didymella lycopersici 214
Diplodia frumenti 268
Diptera 119, 144, 232, 240
Ditylenchus dipsaci 122, 186,
    237, 258

Eggplant mosaic virus 218
Erwinia spp. 224
Erwinia carotovora 249
Erwinia stewartii 268
Erysiphe betae 122
Erysiphe cichoriacearum 168
Erysiphe heraclei 237, 242, 245
Erysiphe pisi 182, 186
Eucalyptus spp. 52

Fabaceae see *Leguminosae*
Foeniculum vulgare 226
Foeniculum vulgare var. *dulce* 226
Fulvia fulva 214
Fusarium spp. 122, 163, 175, 186,
    197, 258, 268
Fusarium oxysporum 142, 172,
    182, 218
Fusarium oxysporum f. sp.
    conglutinans 141
Fusarium oxysporum f. sp.
    lycopersici 204, 214
Fusarium oxysporum f. sp. *melonis* 176

Fusarium oxysporum f. sp. *niveum* 167
Fusarium oxysporum f. sp. *phaseoli* 192
Fusarium oxysporum f. sp. *pisi* 186
Fusarium oxysporum f. sp. *radicis*
    lycopersici 204
Fusarium oxysporum f. sp.
    spinaciae 128
Fusarium oxysporum f. sp.
    vasinfectum 278
Fusarium solani 218, 223, 278
Fusarium solani f. sp. *cucurbitae*
    175, 179
Fusarium solani f. sp. *phaseoli* 192

Galium aparine 84, 146
Gibberella avenacea 138, 245, 249
Gibberella fujikuroi 223, 268
Gibberella fujikuroi var.
    subglutinans 268
Gibberella zeae 268
Glomerella cingulata 214, 278
Glomerella lagenaria 167, 175
Glycine max 181
Gramineae 264–269

Helianthus annuus 38, 53
Hibiscus esculentus 274–278
Hymenoptera 38, 232

Ichneumonidae 41
Ipomoea batatus 222
Itersonilia pastinacea 241, 242

Lactuca indica 129
Lactuca sativa 38, 129–139
Lactuca sativa var. *asparagina* 130
Lactuca sativa var. *capitata* 130
Lactuca sativa var. *crispa* 130
Lactuca sativa var. *longifolia* 130
Lactuca serriola 129, 132
Lagenaria siceraria 180
Leguminosae 3, 32, 55, 100, 181–201
Lepidoptera 240
Lepiium sativum 140
Leptosphaeria maculans 149
Lettuce mosaic virus 131, 136, 137
Lettuce yellow mosaic virus 137

Luffa acutangula 180
Luffa cylindrica 180
Lychnis ringspot virus 122
Lycopersicon cheesmanii 203
Lycopersicon esculentum 202, 203
Lycopersicon lycopersicon 38, 202, 203–214

Macrophomina phaseolina 175
Maize mosaic virus 268
Malvaceae 274–278
Marasmius graminum 268
Marssonina panattoniana 137
Medicago sativa 38, 173
Medicago × *varia* 173
Meligethes spp. 41
Meloidogyne spp. 271
Meloidogyne incognita 204
Melon necrotic spot virus 175
Mormordica charantia 180
Musk melon mosaic virus 175
Musk melon virus 179
Mycosphaerella brassicicola 149
Mycosphaerella piodes 186
Mysus persicae 136

Nasturtium officinale 140, 159–160
Nicotiana tabacum 202
Nosema ceranea 40

Okra leaf curl virus 278
Okra mosaic virus 278
Onion mosaic virus 258
Onion yellow dwarf virus 258
Orobanche spp. 56

Parsley latent virus 245
Pastinaca sativa 226, 238–242
Pea early browning virus 186
Pea enation mosaic virus 182, 186
Pea false leaf roll virus 186
Pea mild virus 186
Pea seed-borne mosaic virus 182, 186
Pemphigus bursarius 54
Penicillium spp. 268
Perenospora destructor 258

Perenospora farinosa f. sp. *betae* 122
Perenospora farinosa f. sp. spinaciae 128
Perenospora viciae 186
Petroselinum crispum 226, 242–245
Phaeoisariopsis griseola 192
Phaeoramularia capsicicola 223
Phaseolus spp. 53, 93, 192
Phaseolus aconitifolius 181
Phaseolus angularis 181
Phaseolus coccineus 28, 181, 191–194
Phaseolus mungo 181
Phaseolus radiatus 181
Phaseolus vulgaris 3, 38, 181, 186–191
Phoma spp. 241, 245
Phoma apiicola 249
Phoma destructiva 214, 223
Phoma lingam 146
Phoma medicaginis var. *pinodella* 186
Phoma rostrupii 237
Phomopsis diachenii 242
Phomopsis vexans 215, 218
Phyllosticta spinaciae 128
Physalis peruviana 202
Phytophthora capsici 219
Phytophthora infestans 204, 214
Phytophthora nicotianae var. parasitica 214
Pimpinella anisum 226
Piper nigrum 218
Pisum arvense 20, 182
Pisum sativum 3, 20, 38, 42, 181
Pisum sativum var. *arvense* 29
Pisum sativum var. *humile* 182
Pisum sativum var. *macrocarpon* 182
Pisum sativum var. *sativum* 29
Plasmodiophora brassicae 142, 149, 156
Pleospora betae 122
Pleospora herbarum 175, 186, 192, 197, 258
Pleospora herbarum f. *lactucum* 137
Poaceae see Gramineae
Populus spp. 52, 54
Potato (Andean) latent virus 224
Potato black ring virus 224
Potato black ringspot virus 224
Potato spindle tuber viroid 214, 224
Potato virus T 224

Potato virus X 224
Potato virus Y 219, 224
Pseudomonas apii 249
Pseudomonas cichorii 137
Pseudomonas cubensis 168
Pseudomonas pseudoalcaligenes ssp.
    citrulli 167
Pseudomonas solanacearum
        204, 223
Pseudomonas syringae 192
Pseudomonas syringae pv. *aptata* 122
Pseudomonas syringae pv.
        lachrymans 172
Pseudomonas syringae pv.
        maculicola 149
Pseudomonas syringae pv.
        phaseolicola 186, 192
Pseudomonas syringae pv. *pisi* 186
Pseudomonas syringae pv. *tomato* 214
Pseudomonas tomato 204
Pseudomonas viridiflava 242
Psophocarpus ringspot virus 200
Psophocarpus tetragonobolu 181,
        199–200
Puccinia allii 258
Pyrenochaeta lycopersici 204

Ramularia beticola 122
Raphanus landra 152
Raphanus maritimus 152
Raphanus raphanistrum 152
Raphanus rostatus 152
Raphanus sativus 140, 152–155
Raphanus sativus var. *mougri* 152
Raphanus sativus var. *niger* 152, 153
Raphanus sativus var. *oleifera* 152
Raphanus sativus var. *radicula* 152, 154
Raspberry ringspot virus 122
Rhizoctonia petroselini 245
Rhizoctonia solani 128, 138, 149,
        186, 192, 214, 218, 223, 278
Rorippa microphylla 159
Rorippa nasturtium-aquaticum 140,
        159–160
Runner bean mosaic virus 192

Salix spp. 52, 54
Sclerotinia cepivorum 258

Sclerotinia sclerotiorum 137, 149,
        186, 192, 218, 223
Sclerotium rolfsii 137
Sclerphthora macrospore 268
Sclerphthora maydis 268
Sechium edule 180
Septoria apiicola 246, 249
Septoria lactucae 137
Septoria petroselini 245
Sinapis alba 140
Solanaceae 48, 80, 202–225
Solanum aethiopicum 202
Solanum macrocarpon 202
Solanum melongena 202, 214–218
Solanum muricatum 202
Solanum quitoense 202
Solanum tuberosum 202, 222–224
Sorghum spp. 53
Sphaerotheca fulginea 168
Spinach latent virus 128
Spinacia oleracea 38, 44, 116, 125
Squash mosaic virus 167, 175, 179
Stemphyllum spp. 204
Strawberry latent ringspot virus 242,
        245, 249, 274
Sugar cane mosaic 268

Tamarix spp. 52
Taraxicum officinale 129
Telfairia occidentalis 162
Tobacco mosaic virus 204, 212, 214,
        215, 223
Tobacco mosaic virus, streak strain 214
Tobacco mosaic virus, tomato mosaic
        strain 214
Tobacco ringspot virus 137, 175,
        214, 224
Tobamovirus 219
Tomato black ring virus 122, 137
Tomato yellow leaf curl virus 204
Tragopogon porrifolius 129
Trichosanthes cucumerina 180
Trigonella foenum-graecum 181
Turnip mosaic virus 150

Umbelliferae 226–250
Urocystis cepulae 258
Uromyces viciae-fabae 197

Ustilaginoidea virens 268
Ustilago zeae 268

Varroa spp. 40
Verticillium spp. 128, 204
Verticillium albo-atrum 218, 249
Verticillium dahliae 128, 214, 218
Vicia faba 29, 181, 194–197
Vigna subterranean 197
Vigna unguiculata 181
Voandzeia subterranea 181, 197–199
Voandzeia subterranea var.
    spontanea 198
Voandzeia subterranea var.
    subterranea 198

Xanthomonas campestris pv.
    campestris 149
Xanthomonas campestris pv.
    carotae 237

Xanthomonas campestris pv.
    cucurbitae 179
Xanthomonas campestris pv.
    phaseoli 192
Xanthomonas campestris pv.
    vesicatoria 223
Xanthomonas rubifaciens 186

Zea spp. 100
Zea mays 38, 53, 93
Zea mays var. *amylacea* 264
Zea mays var. *ceratina* 264
Zea mays var. *erythrolepis* 264
Zea mays var. *everata* 264
Zea mays var. *indentata* 264
Zea mays var. *indurata* 264
Zea mays var. *japonica* 264
Zea mays var. *praecox* 264
Zea mays var. *saccharata* 264–269
Zea mays var. *sugosa* 264
Zea mays var. *tunica* 264